NANOTECHNOLOGY SCIENCE AND TECHNOLOGY

NANOSTRUCTURED MATERIALS: CLASSIFICATION, PROPERTIES AND FABRICATION

NANOTECHNOLOGY SCIENCE AND TECHNOLOGY

Additional books in this series can be found on Nova's website under the Series tab.

Additional E-books in this series can be found on Nova's website under the E-books tab.

Nanotechnology Science and Technology

Nanostructured Materials: Classification, Properties and Fabrication

Anees A. Ansari
M. Naziruddin Khan
M. Alhoshan
A.S. Aldwayyan
and
M.S. Alsalhi

Nova Science Publishers, Inc.
New York

For permission to use material from this book please contact us:
Telephone 631-231-7269; Fax 631-231-8175
Web Site: http://www.novapublishers.com

LIBRARY OF CONGRESS CATALOGING-IN-PUBLICATION DATA

Nanostructured materials : classification, properties, and fabrication /
authors, Anees A. Ansari ... [et al.].
p. cm.
Includes index.
ISBN 978-1-61668-763-2 (softcover)
1. Nanostructured materials. 2. Metallic oxides. I. Ansari, Anees A.
TA418.9.N35N3534 2010
620.1'1--dc22
2010025440

Published by Nova Science Publishers, Inc. † New York

CONTENTS

Preface		vii
Chapter 1	Introduction	**1**
Chapter 2	Classification of Nanostructured	**5**
Chapter 3	Properties of Nanostructured Materials	**7**
Chapter 4	Preparation of Nanostructured Materials	**25**
Chapter 5	Characterization of Nanostructured Materials	**45**
Chapter 6	Nano-Structured Materials for Biomedical Applications	**61**
Chapter 7	Conclusion and Future Prospects	**85**
References		**87**
Index		**117**

PREFACE

The recent surge of interest in the nanotechnology has significant expanded the breadth and depth of nanostructured materials. The nanoscale grain size can facilitate important advances in the development of nanodevices and nanomedicine to reduce the problems of disease to problems of molecular science, and are creating new opportunities for treating and curing disease. Such advances have led to a new discipline of nanotechnology and lead to a variety of approaches for relieving suffering and prolonging life. This new book presents and discusses the uniqueness of chemistry, structure, response and dynamics of nanostructured materials.

Chapter 1

INTRODUCTION

In the rapidly changing scientific world, nanotechnology is one of the most emerging field especially in material and biomedical sciences for technology applications in the past two decades [1-9]. Recent surge of interest in the nanotechnology has significant expanded the breadth and depth of nanostructured materials [10]. The nanoscale grain size can facilitate important advances in the development of nanodevices and nanomedicine to reduce the problems of disease to problems of molecular science, and are creating new opportunities for treating and curing disease [11-26]. Such advances have led to a new discipline of nanotechnology and lead to a variety of approaches for relieving suffering and prolonging life. Nanotechnology defines as the creation of functional materials, devices and systems through control of matter at the range of 1-100 nm scale [27]. The chemical and physical properties of nanostructured materials can be significantly different from those of atomic, molecular and bulk materials of the same composition [1-4, 27-30]. The uniqueness of chemistry, structure, response and dynamics of the nanostructures is the indispensable motivation for the study of this class of materials. Suitable control of the properties of nanometer scale structures can lead to new science as well as new products, devices and technologies [28-30]. The development of new sophisticated tools like STM, HRTEM, SESEM, AFM, XRD etc., to observe, measure and manipulate processes at the nano-scale level gave a breakthrough to the nanotechnology. The term nanotechnology was first used by the Japanese researcher Taniguchi in 1974 when he referred to the ability to engineer materials at the nanometer scale [31]. The main idea behind this terminology was the miniaturization in the electronics industry. Even as early as 1970s a huge number of nanostructures were created as small as 40-70 nm using electron beam lithography. The term nanomaterials covers materials in one dimension (quantum dots and thin films etc.) [27], two dimension (nanofibers, nanowires, nanotubes,etc.), and three dimension (nanoparticles/nanopowders, nanocapsules, fullerenes, dendrimers, molecular electronics, nanostructured materials, nanoporous materials etc.). Due to their small grain size, these materials exhibit unique mechanical, chemical, physical, thermal, electrical, optical, magnetic, biological and also specific high surface area properties, which cannot be found in their bulk counterpart turn define them as nanostructures, nanoelectronics, nanophotonics, nanobiomaterials, nanobioactivators, nanobiolables, etc. Therefore, many researchers have been devoted to developed new synthesis routes, novel characterization methodologies and their application in bionanotechnology. Until now, a wide variety of nanostructured materials and devices with new capabilities have been generated employed in the literature some of them are polymeric materials such as (polyaniline, polypyrrole, polythiophene, polyethylene glycol, chitosan, polysaccharides etc.) [32-43], CNT(multi-walled and single walled carbon nanotubes) etc.) [44-49], metallic nanoaprticles (Ag, Au, Ge, Mo, Si, Pt, Pd and Ru etc.) [50-74], semiconductor metal oxides (Al_2O_3,[75,76] CeO_2[77-80], CdO[81,82], CoO,Co_3O_4[83], CuO,Cu_2O [84,85], FeO, Fe_2O_3, Fe_3O_4[86-112], Ga_2O_3 [113], Hf_2O_3[114], In_2O_3[115-120], Ln_2O_3[121] MgO [122,123], MoO_3[124,125], MnO_2, Mn_2O_3, Mn_3O_4[126-131] Nb_2O_5[132,133] NiO[134,135], PbO_2[136,137] Pr_2O_3[138-140] Sb_2O_3[141,142], SiO_2[143-147], SnO_2[148-155] V_2O_5[156-158], WO_3[159-161] TiO_2[162-175] ZnO[176-186], ZrO_2[187-194], Ln_2O_3[195-205] (Ln = Y, La-Lu), YVO_4[206-214]), fluorescent metal nanoparticles (LnF_3, LnF_3:Ln, YF_3, YF_3:Ln [215-226], $NaYF_4$, $NaYF_4$:Ln $NaLnF_4$, $NaLnF_4$:Ln[227-248], $LnPO_4$, $LnPO_4$:Ln[249-258], (Ln = La-Lu)), semiconductor quantum dots (AgS, AlN, CdS, CdSe, CdTe, CuS, GaN, GaP, PbS, PbTe, ZnS, ZnSe, ZnTe etc. [259-280]) and their hybrid derivative nanoparticles with different properties for wide applications in the field of biomedical sciences. These size and structured dependent nanostructured materials offer excellent prospects in a wide variety of other practical applications, such as transparent electrodes, catalysis, solar energy conversion, field emission, photonic devices, drug delivery, biosensors(DNA, optical, chemical, gas and immunosensors) bio-labeling, biomarker, magnetic resonance imaging (MRI), laser technology, tissue engineering and in forensic sciences[281-295].

Due to specific surface properties of nanostructured materials some investigators have been exploited their applications in technology development and biomedical sciences. Noticed that the diameter of nano-structured materials, surface condition, crystal structure and its quality i.e., chemical composition, crystallographic orientation are key important parameters that influence the surface properties of the nanomaterials [296]. Moreover, the conductance of nanostructured materials strongly depends on their crystalline structure. For instance, in the case of perfect crystalline Si nanowires having four atoms per unit cell, generally three conductance channels are found. One- or two-atom defect, either by addition or removal of one or two atoms may disrupt the number of such conductance channel and may cause variation in the conductance. Observation results revealed that change in the surface conditions of the nanowires can cause remarkable change in the transport, electrical, mechanical and catalytic behavior.

The manipulation of nanometer-scale structures can lead to new sciences as well as new devices and technologies. In the last two decade, some new synthesis methodologies have been employed for the preparation of nanocrystals of metals, semiconductors, quantum dots and magnetic nanomaterials using various chemical and physical methods. Detail on various preparation techniques are being discussed with their classification, novel properties, synthesis methodologies, characterization and applications that are likely to fuel research into the coming decades. And other related nanoparticle syntheses by various physical and chemical synthesis process is discussed in this chapter. The goal of this chapter is to summaries the classification, properties, fabrication process, characterization technique of nanostructured materials and their advancement in the development of biomedical sciences.

Chapter 2

2. CLASSIFICATION OF NANOSTRUCTURED

Nanostructured describe the intermediate size between molecular and microscopic (micrometer-sized) structures. They can be divided according to their crystal symmetry and dimensionality, which highly affects the functionality of the correspondence parameter. Further the dimensionalities of nanomaterials are commonly classified into zero dimensional (dots or mostly spherical nanoparticles), one dimensional (nanowires, nanotubes and nanobelts) or two-dimensional (thin films). These categories refer to the number of dimensions in which the material is outside the nano regime. A thin film, for example, consists of large expanses of material in both in-plane directions and nanosized only in its thickness; therefore it is termed as two-dimensional (2D) nanomaterial. Due to the effects of size and dimensionality of a nanomaterial can alter its properties, such that a thin film may behave very differently from a nanowire with the same chemistry. These distinctions therefore are not arbitrary, but important in distinguishing the materials according to unique characteristics. Figure 1 shows the geometrical differences between the classes of nanostructures.

Nanodiscs, nanoplates, nanosheets, and nanomembranes are two-dimensional (2-D) nanostructures: the x and y directions are fully extended with only the thickness along the z direction between 1 and 100 nm of the polygon-shaped surfaces. Nanorods, nanowires and nanotubes are examples of one-dimensional (1-D) nanostructures. They can be hollow tubes or cylinders with polygon shapes of nanoscale diameter while the length of the nanostructures increases. Nanostructures with zero dimensions (0-D) on the nanoscale are classified as nanoparticles, each spatial dimension having a scale within 100 nm. The shapes can be spheres, isotropic spheres, icosahedrons, octahedrons, cubes, or tetrahedrons.

The potential of these materials use for structural, commercial, environmental, electronic, and biological applications, have lately sparked their interest and imaginations in all science disciplines, and even the general public. While certainly improved synthesis techniques have helped move 1D nanomaterial research forward, the driving force behind their rapid development is a new appreciation for their unique properties, driven by the need to utilize these properties in highly demanding applications (Table 1).

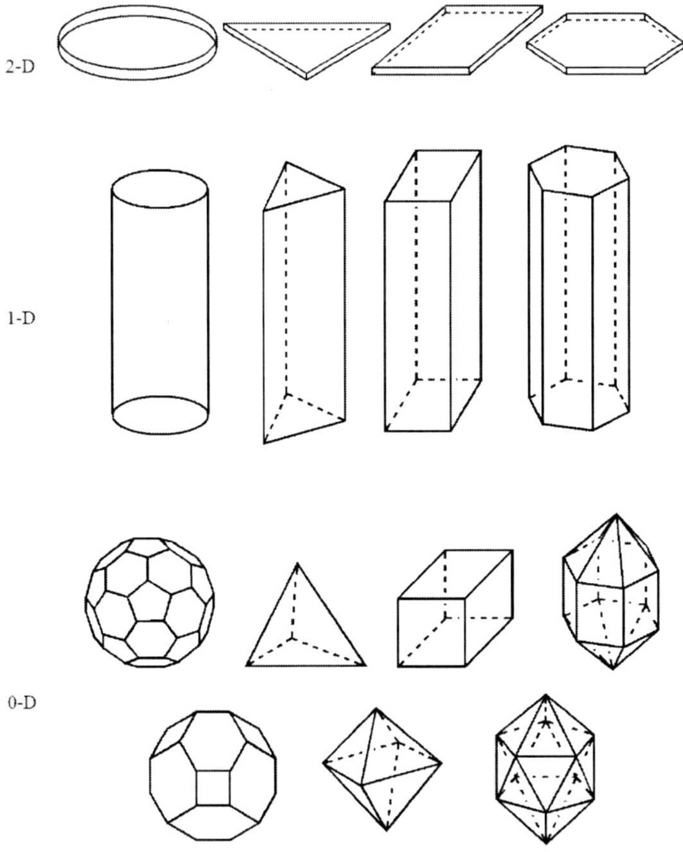

Figure 1. Geometric shapes of basic nanostructures.

Chapter 3

3. PROPERTIES OF NANOSTRUCTURED MATERIALS

Materials in the micrometer scale usually exhibit similar physical properties as that of bulk form; however, materials in the nanometer scale may exhibit distinctively different physical properties. When the dimensions of materials go down to nanometer scale, some remarkable specific physical, chemical and biological properties due to increased surface area, smaller particle size, reduced number of free electron, and quantum confinement effect would be developed. Size effect has a dramatic impact on structural, thermodynamic, electronic, spectroscopic, electromagnetic and chemical properties of nanomaterials. Size affects the structure, melting point and the electronic absorption of nanomaterials such as that of ZnS and ZnO. For example, semiconductor nanocrystals are zero-dimensional quantum dots, in which the spatial distribution of the excited electron-hole pairs are confined within a small volume, resulting in enhanced non-linear optical properties. Due to the small grain size, the electronic properties of a semiconductor nanocrystal are significantly affected by the transport of a single electron, giving the possibility of producing single electron devices.

Semiconductor crystals exhibit a broad spectrum of strong size dependent properties for particle sizes below the Bohr radius of their bulk exciton. The size dependence of the nanocrystal electronic energy level has attracted special interest. It allows one to tailor the materials optical and electronic properties by controlling particle size and many potential new applications such as nanocrystal bio-tags, quantum dot lasers, polymer-based nanocrystal solar cells and single electron transistors have been suggested.

Table 1. Summery of the siz of different nanomaterials

Morphology	Typical size of nanomaterials	Materials
Nanocrystals and clusters (quantum dots)	Diameter: 1~10 nm	Metals, semiconductors, magnetic materials
Other nanoparticles	Diameter: 1~100 nm	Ceramic oxides, metal oxides etc.
Nanowires	Diameter: 1~100 nm Length: Up to several micrometers	Metals, oxides, sulfides, nitrides etc.
Nanotubes	Diameter: 1~100 nm Length: Up to several micrometers	Carbon, layered metal chalcogenides etc.
Nanorods	Diameter: 1~100 nm Length: Up to several micrometers	Metals, oxides, sulfides, nitrides etc.
Nanoporous solids	Pore diameter: 0.5~10 nm	Zeolites, phosphates etc.
2-Dimensional arrays (of nanoparticles)	Several nanometers to several micrometers	Metals, semiconductors, magnetic materials
Surface and thin films	Thickness: 1-1000 nm	A variety of material, such as Ag film
3-Dimensional structures (superlattices)	Several nanometers in the three dimensions	Metals, semiconductors, magnetic materials

The concept of quantum confinement describes the effects of reduced dimensions on the electronic properties of nanostructures. If the size of the nanoparticle becomes similar to the wavelength of the charge carriers in the valence and conduction band, the particle can be described as a potential well where waves are reflected at the potential barrier and spatially confined. As a result, the electronic structure of the crystal is altered from continuous bands to discrete levels and the continuous optical transitions between the electronic bands become discrete and size-dependent as well. The basic idea behind quantum confinement can be visualized as a classical particle in a box problem. For a free particle with an effective mass of m^* confined in a crystal with a one dimension barrier at a distance of L, the movement of the particle can be described with the Schröedinger equation:

$$\frac{\partial^2 \Psi(x)}{\partial x^2} = -\frac{2m^* E}{(h/2\pi)^2}\Psi(x) \qquad \text{(Eq. 1)}$$

where $\Psi(x)$ is the wave function and E is the energy of the particle. This equation can be solved by using the boundary conditions that $\Psi(x) = 0$ at $x = 0$ and $x = L$:

$$\text{for} \quad E_n = \frac{\hbar^2 k_n^2}{2m^*} = \frac{n^2 \hbar^2 \pi^2}{2m^* L^2} \quad n = 1,2,3. \qquad \text{(Eq. 2)}$$

where k_n is the allowed wave factors and $k_n = n\,\pi / L$

When this simplified solution is extended to a semiconductor nanocrystal with a size close to the Bohr radius a_B, the resulting expression for the energy shift from the ground state of the corresponding bulk materials is:

$$\Delta E_n = n^2 \hbar^2 \pi^2 / 2m^* a^2 \qquad \text{(Eq. 3)}$$

where a is the size of the nanocrystal, m^* is the reduced effective exciton mass which is defined by $1/m^* = 1/m_e^* + 1/m_h^*$.$me^*$ and mh^* are the effective mass of electron and hole of the material, respectively. The Bohr radius of a semiconductor can be obtained from the dielectric constant ε and the reduced exciton mass m^* : $a_B = \varepsilon h^2 / m^* e^2$.

Thus, the band gap of the material can be engineered to a desired level by changing the size of the nanocrystals:

$$E_{g,\ nanocrystal} = E_{g,\ bulk} + \Delta E = E_{g,bulk} + \hbar^2 p^2 / 2m^* a^2 \qquad \text{(Eq.4)}$$

The band gap energy increases with a decrease in the size of the nanocrystals. Since the optical transition is directly related to the band gap, it is possible to tune the absorption and photoluminescence wavelength of the nanocrystals just by manipulating their size.

There are numerous properties of metal oxides that are affected by decreasing the grain size within the material. Some of the most important properties of nanophase metal oxides are discussed here. Mechanical, electronic, magnetic properties and catalytic of nanophase materials have important applications. Also, the transport properties play an important role in technological development. These properties make nanophase materials attractive to study both for fundamental aspects as well as for potential novel applications (Figure 2).

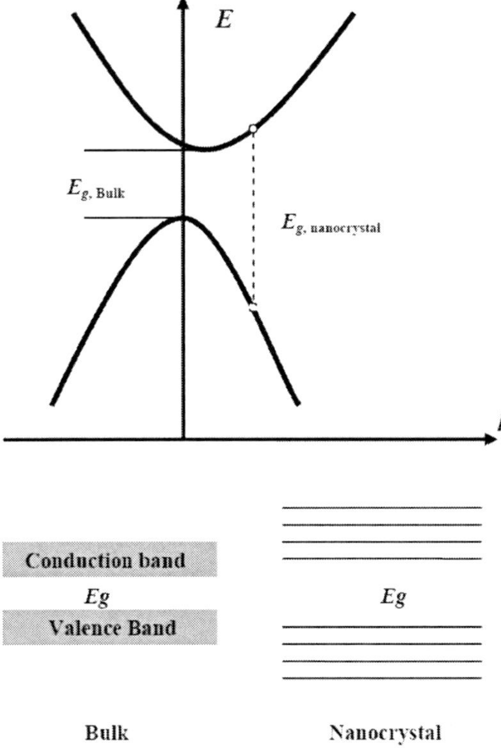

Figure 2. Illustration of a semiconductor band structure showing shift of energy band gap.

3.1. Mechanical Properties

In metal oxides formation, grain size reduction can yield improvements in strength and hardness. Grain size reduction increases the strength and hardness due to new grain boundaries, which act as effective barriers to dislocation motion. However, it may affect other mechanical properties negatively, such as creep rate and ductility. On the other hand, in materials that are conventionally quite strong but very brittle, such as intermetallic compounds and ceramics (ZnO, CeO_2, ZrO_2, In_2O_3, WO_3, TiO_2 and SnO_2), enhanced ductility from grain size reduction, through the increased grain boundary sliding can be considered favorable.

3.1.1. Hardness

While the hardness of materials clearly increases as their grain size is reduced into the nanophase regime (Figure 3) that is dependent in the variation of the grain size. Increased hardness with decreased grain size conflicts with the softening of nanophase materials seen in samples annealed to increase their grain size. Inter metallic alloys show initial hardening with decreasing grain size, but at further reduced grain sizes two different cases with either hardening at a reduced slope or softening can be separated. Although hardening in nanophase materials is analogous to Hall-Petch strengthening it is considered to result from different mechanisms. The grain sizes in ultrafine particles are smaller than the critical bowing length for Frank-Read dislocation sources needed to operate at the stresses involved and smaller than the normal spacing between dislocations in a pile-up. The presence of sample porosity, flaws, or contamination from synthesis and processing could influence the available hardness results, as could the nature of the grain boundaries and their state of relaxation.

3.1.2. Thermal Stability

Thermal stability of nanocrystalline materials shows basically that nanocrystalline metals exhibit grain growth at relatively low annealing temperatures. This is to be expected because of the large energy stored in the material as a result of the large volume fraction of grain boundaries. The general observation emerging from these studies is that for metals with an equilibrium melting temperature (T_m) less than approximately 873K and with starting grain size of 10nm or less, doubling of the grain size takes place in approximately 24 hours at about ambient temperature. However, for metals with high T_m, the stability against grain growth seems to be improved. Gleiter [297] reported that nanocrystalline iron with a starting grain size of about 10nm, is thermally stable up to 473K. The grain size increases by a factor of 5 after annealing for 10 hours at 673K. Annealing at 773K for 1 hour transformed the material into the conventional polycrystalline state. Figure 4 shows the grain size as a function of the annealing time for nanocrystalline Ni-P at various temperatures. The results indicate no grain growth for up to 10 hours at the lowest annealing temperature (473K). However, at the intermediate temperature, 573 and 623K, the grain size initially increased rapidly by a factor of about 2 to 3 before stabilizing at 15 and 22nm, respectively.

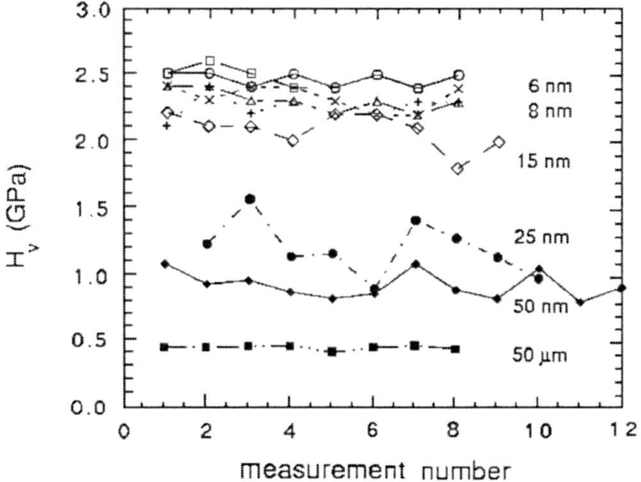

Figure 3. Microhardness (Vickers) Measurements across nanophase Cu samples ranging in grain size from 6 to 50 nm and compared with conventional 50 μm grain size Cu.

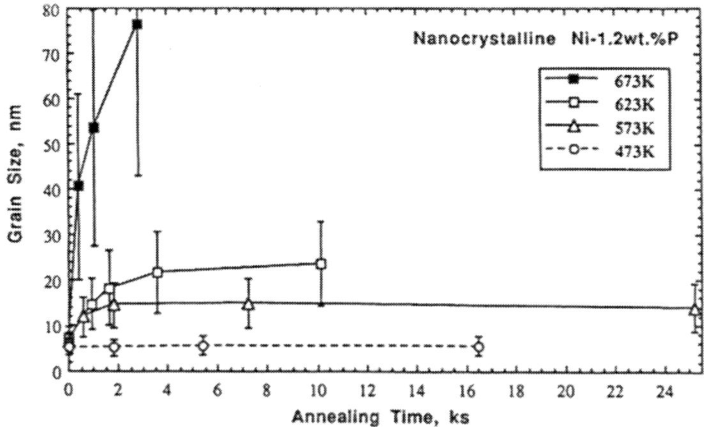

Figure 4. Grain size as a function of the annealing time for nanocrystalline Ni-P at various temperatures [From reference 298].

3.1.3. Tensile Strength and Fracture

Face-centered cubic metals tested in tension have exhibited similar improvements in strength as those seen in their hardness behavior, but they have shown limited ductility. The limited levels of ductility exhibited by nanophase metals may arise because of difficulties in creating, multiplying, and moving dislocations, but may as well relate to the presence of significant flaw populations in these materials. The results to date on the fracture properties of nanophase materials have been limited in scope and hindered by the presence of porosity or interfacial phases [299]. Conventionally brittle ceramics have been observed to become ductile, permitting large plastic deformations at low temperature if the ceramics are generated in the nanocrystalline form. Doping of nanocrystalline ceramics has been demonstrated to reduce grain growth dramatically. Grain boundary sliding, grain rotation and grain shape accommodation by diffusion processes seem to play a crucial role in the deformation of nanocrystalline ceramics, i.e. processes that are typical for superplasticity [300].

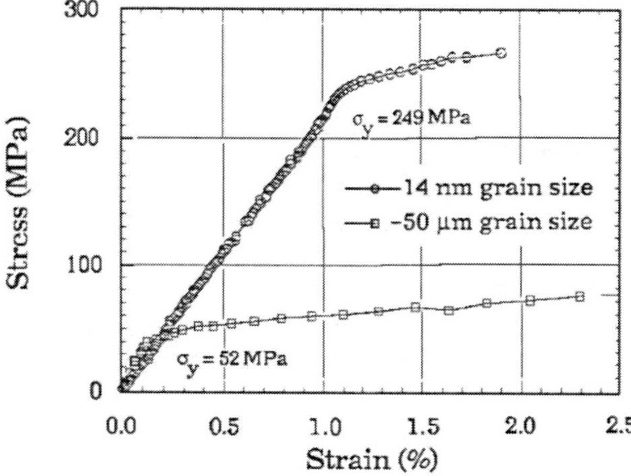

Figure 5. A stress-strain curve for nanophase Pd compared with that for a coarse-grained Pd sample [301].

It appears that the increasing hardness and strength in nanophase metals with decreasing grain size indicates a result of diminishing dislocation activity and increasing grain boundary density. Figure 5 shows, the frequency of dislocation activity decreases and that of the grain boundary sliding increases [301].

3.2. Transport Properties

The solid oxide fuel cell is a candidate for efficient electric power generation systems. Solid electrolytes used in such cells have been increased with investigation for many years due to their suitability as ionically conductive materials at high temperatures, such as solid oxide fuel cells and oxygen sensors. A considerable amount of research is being carried out in order to develop materials suitable for applications in the medium temperature range [302]. These solid electrolytes are mostly fluorite structure. Open structure is required for the material to have high ionic conductivity. The IVB group oxides ThO_2, CeO_2, PrO_2, UO_2 and PuO_2 have the fluorite structure and by doping it is possible to stabilize this structure for ZrO_2 and HfO_2. The addition of dopants gives rise to the creation of oxygen vacancies responsible for the ionic conductivity in these oxides [303].

3.2.1. Ionic and Electronic Conductivity

There are two important ways in which materials can conduct electrical current. Both electrons and ions can carry electric charge. For metals, the electronic conductivity dominates while for many ceramic oxides ionic conductivity is the major contributor that may exceed electronic conductivity by several orders of magnitude. In metals the non-localized electrons carry the charge. The ionic conductivity is associated with ion motion. Lattice defects are of great importance for ionic conductivity [304]. Diffusion and ionic conductivity are related through the Nerst-Einstein equation.

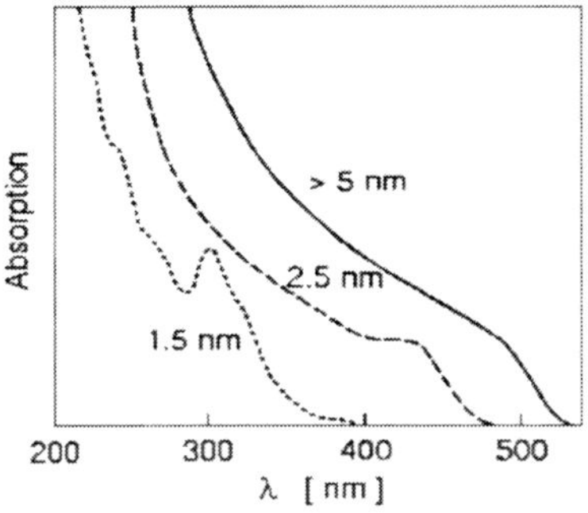

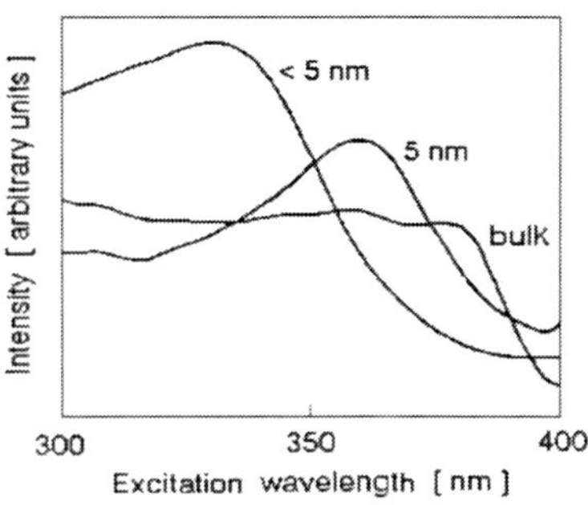

Figure 6. Absorption spectrum of CdS particles in aqueous solution as a function of particle size (left). Photoluminescence spectra of nanocrystalline ZnO with different crystal sizes in comparison with the bulk material (right) [308].

By doping cerium oxide, it becomes an anionic conductor at high oxygen pressures. Lower valent cations in the lattice make the formation of anion vacancies energetically more favorable than the formation of electron holes. At intermediate oxygen pressures n-type conductivity interferes [303].

3.2.2. Effects of Particle Size and Doping on Electrical Conductivity

Transport properties of ceramics, in semi- and ionic conductors, are often limited by grain boundaries. The defect and transport properties of nanocrystalline oxides are unique in two principal aspects. Grain boundary impedance is greatly reduced due to size dependent impurity segregation [305]. In varistors as an example this gives rise to useful electrical barriers. Defect thermodynamics dominated by interfaces are obtained in a bulk material when particle size is in the nanometer regime. The unusual defect thermodynamics of the nanocrystals are attributed to interfacial reduction [306]. The heat of reduction of nanophase ceria is lowered by more than 2.4 eV per oxygen vacancy compared to coarse particles. This reduction occurs at suitably low-energy to dominate the nonstoichiometry and electrical conductivity of the material as whole.

Diffusion usually takes place by the movement of ions to neighboring vacancies. Oxygen atoms are located in the oxygen interstitials on non-fluorite sites. For stoichiometric compounds, the vacancy concentration and ionic conductivity are very small, but suitable doping may increase both. Also, smaller particle size increases the non-stoichiometry of a material. The oxygen ion conductivity of doped cerium oxide with a constant doping level has been found to be dependent on the lattice constant of the compound. Low doping changes the conductivity of ceria from an ionic to an electronic conductor [307].

3.3. Electronic Properties

For small particles, the wave functions of electrons and holes are confined to the particle volume. If the particle size becomes comparable to or smaller than the de Broglie wavelength of the charge carriers, the confinement increases the energy required for creating an electron/hole pair. This increase shifts the absorption/luminescence spectra towards shorter wavelengths (so called "blue shift" [308]) as shown in Figure 6. In direct gap semiconductors, the band gap increases with decreasing particle size and the excited electronic states become discrete with high oscillator strength. The optical absorption spectrum of γ-Fe_2O_3 is red shifted, which can be explained by the lattice strain in small particles [309]. In a polymer matrix γ-Fe_2O_3 shows optical transparency in the visible range.

3.3.1. Indirect Semiconductors

Nanocrystalline Si and porous Si are indirect semiconductors where the absorption involves both a photon and a phonon because the conduction and valence bands are widely separated in k space [310]. Figure 7 shows the differences in optical absorption and optical transition in direct and indirect gap semiconductors.

A direct photon transition at the minimum gap cannot satisfy the requirement of conservation of energy, because photon wave vectors are negligible at the energy range of interest. But if a phonon of wave vector K and frequency W is created in the process, then the excitation can occur. Several vibronic interactions should increase when wave functions become more compact in nanocrystals. Thus, the vibronic luminescence rate as well as the electronic luminescence rate should increase with decreasing size in nanocrystals. For sizes larger than roughly 15 Å, the vibronic luminescence dominates. Although these nanoparticles have high luminescence yields, the oscillator strengths of nanocrystalline Si and porous Si are not markedly increased with respect to bulk Si [310-312]. In silicon the quantum size effect is primarily kinetic, while in the conventional semiconductors the effect is more spectroscopic [313]. Silicon luminescence increases because spatial confinement keeps the electron and hole superimposed and surface nonradiative rates are also extremely slows.

3.4. Magnetic Properties

3.4.1. Magnetism of Multi and Single Domain Particles

A crystal will spontaneously break up into a number of domains in order to reduce the large magnetostatic of single domain particles. Dipolar energy can be reduced by dividing a magnetic specimen into uniformly magnetized domains of macroscopic size, whose magnetization vectors point in widely different directions. Such subdivision is paid for in exchange energy, for the spins near the boundary of a domain will experience unfavorable exchange interactions with the nearby spins in the neighboring domain. Only the spins near the boundaries will have higher exchange energies but the magnetic dipolar energy drops for every spin when domains are formed [314]. When the crystal size is reduced below a critical value Lc of a few hundred Å a single crystal will become a single domain. In the single domain particles it is not energetically favorable to get magnetization reversed in the particle so no walls are formed.13

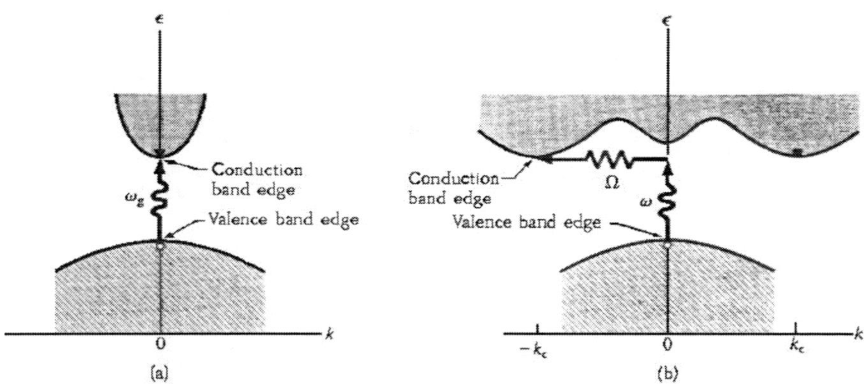

Figure 7. Optical absorption in pure insulators with direct and indirect gaps at absolute zero [310].

The extrinsic magnetic properties of particles depend strongly upon their shape and size. This result was obtained when magnetostatic, exchange and domain wall energies were considered. Interactions between particles also affect the critical size. Among magnetic properties coercivity Hc shows a marked size effect. Saturation magnetization on the other hand is not dependent on the particle size [315].

3.4.2. Superparamagnetism

Superparamagnetic particles posses a magnetic moment that may be many orders of magnitude greater than that of a paramagnetic atom $\mu = MsV$ and, if a field is applied, the field will tend to align the moments of the particles. Thermal energy on the other hand will tend to misalign them. This is just like the normal paramagnetic behavior, just with an exceptionally large magnetic moment. In very small particles random thermal forces are large enough to cause the magnetization direction to reverse spontaneously between the easy directions [316]. The average time between reversals is an exponential function of the ratio of the particle volume to absolute temperature. A particle will spontaneously reverse its magnetization even in the absence of an applied field when the energy barrier to rotation is about 25kT [317]. When the field is turned off the initial magnetization will begin to decrease due to the thermal energy. The rate of decrease is proportional to the magnetization existing at the time and to the Boltzmann factor $e^{KV/kBT}$. Particles that exhibit super paramagnetic behavior have a large saturation magnetization but no remanence or coercivity. Hysteresis will appear and super-paramagnetism disappears, when particles of a certain size are cooled to a particular temperature, or when the particle size, at constant temperature, increases beyond a particular diameter. For superparamagnetic particles magnetization curves measured at different temperatures superimpose when M is plotted as a function of H/T. Since the direction of magnetization in SP particles is changing, the particles cannot be used for magnetic recording. Potential applications of nanoscale magnetic particles are in color imaging, ferro fluids and magnetic refrigeration.

Nano size metallic particles may be so reactive that they undergo surface oxidation [318-321]. Magnetism of these particles can be explained with a core-shell structure, where the core consists of metallic iron and the shell is composed of iron oxides. The magnetic properties depend strongly on particle size and the amount of oxidation. The saturation magnetization is determined from M vs. $1/H^2$ plots by extrapolating the value of magnetization to infinite fields. Ms was found to increase with increasing particle size. The decrease in magnetization with decreasing particle size is related to the higher surface to volume ratio in the smaller particles resulting in a much higher contribution from the surface oxide layer.

The coercivity of the particles is found to depend strongly on particle size [318,319]. Hc decreases as a function of temperature and decreasing particle size makes the difference in coercivity larger by increasing the coercivity at low temperature and dropping to zero at lower temperatures. Larger surface to volume ratio could result in a higher effective anisotropy with large contributions, from the surface and interface anisotropy. In addition to the usual magnetocrystalline and magnetostatic anisotropies, often there are strong influences from the anisotropies due to strains, surface and magnetic interactions coming from the other particles. Experimentally it has been found that K is one or two orders of magnitude larger for Fe, Fe-oxides and Fe-SiO_2 in the nanosize particles than in bulk [318,319]. The large effective anisotropy is thought to partly originate from the shell-type particle morphology with strong shell-core interactions and partly from the large surface effects expected in ultrafine particles.

3.5. Catalytic Properties

Studies on catalytic properties are predominantly based on single crystal surfaces of metal in vacuum. There is very little information on catalytic processes involving oxides. Catalytic reactions involve the following general surface chemical steps [322].

1. Adsorption of the reactant molecules at surface sites
2. Bond breaking of the adsorbates
3. Rearrangement of the adsorbed reaction intermediates
4. Desorption of the molecules

The influence on decreasing the particle size at different stages in catalysis is discussed in the following sections. Materials become more stepped and rough as the particles become smaller in size. Doping gives rise to anisotropic surfaces.

3.5.1. Particle Size Dependence of Surface Processes

Several surface processes are dependent on the surface structure and particle size. It has been shown that some catalytic reactions are structure insensitive, and only one active surface metal atom is needed. For structure-sensitive reactions several adjacent active surface atoms are needed. Reactions involving M-O bond formation or breaking are usually structure insensitive, while reactions involving O-O bond breaking/formation are usually structure sensitive. For the structure-sensitive reactions particle size and other properties can be of importance. Both catalytic activity and selectivity is dependent on particle size [323].

3.5.2. Adsorption

All catalytic reactions proceed through adsorption. The adsorbate atoms are trapped at sites where the well depth of the attractive surface potential is higher than the kinetic energy of the atoms. The sticking probability of atoms is defined as the rate of adsorption divided by the rate of collision, and it is higher for more open and rough surfaces than on flat ones. This is due to the higher heat of adsorption at a kink or step edge site.

3.5.3. Diffusion and Desorption

Diffusion is very fast on catalyst surfaces and requires a much lower activation energy than desorption. Diffusion is anisotropic since diffusion rates parallel to steps are greater than diffusion rates perpendicular to steps. The presence of co-adsorbates influences surface diffusion markedly. Elements that decrease the melting point of the substrate cause an increase in surface diffusion rates and a decrease in activation energies for diffusion in general. The desorption of atoms from kink and step sites requires a much higher energy than from flat surfaces. The change of surface from flat to stepped and kinked gives the material a wider range of desorption/adsorption energies.

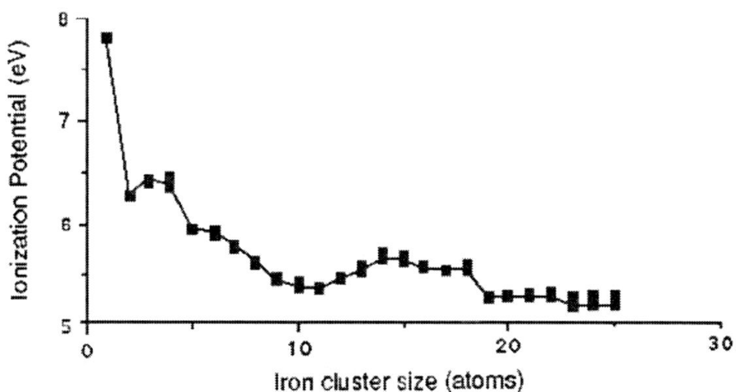

Figure 8. The ionization potential of iron clusters as a function of cluster size [324].

3.5.4. Work Function

The work function is the ionization potential required to eject the most loosely bound electrons into vacuum at 0 K. The work function depends on both the surface roughness and the particle size of the material. It is easier to ionize metals when step density increases [324]. The effect of particle size on the work function is less studied but there is evidence that the ionization potential increases with decreasing particle size (Figure 8). It is not yet known whether these changes are due to changes in cluster structure or some other effect.

3.5.5. Bond Breaking

Bond breaking of the adsorbate molecules is the main function of catalysts. Bonding of adsorbate to metal generally increases from left to right in the periodic table. The d-electron back bonding depends on the degree of filling of the antibonding states. This means that early transition metals with fewer d-electrons form stronger chemical bonds. The heat of adsorption correlates with the heats of formation of the corresponding oxides.

Molecules dissociate on more open and atomically rough surfaces at lower temperatures than on flat, and close-packed surfaces. The heat of adsorption is higher at defect sites on a surface. Defect sites, surface roughness and low packing density of surfaces give rise to higher charge densities near the Fermi level. The work function of these sites is lower and the density of filled electronic states is higher. They are likely to have more adsorbate-induced restructuring. These factors can give rise to enhanced reactivity and bond strength and lead to surface-structure sensitivity of the adsorbate bond.

3.5.6. Charge Transfer

Most of the catalytic reactions involve charge transfer of either electrons or protons. Electron transfer capability is described as a site being able to receive a pair of electrons (Lewis base) or having a free pair of electrons that can be transferred to the adsorbate (Lewis acid). Acidity of metal ions of equal radius increases with increasing charge. The binding energies of the charge-transfer complexes determine the strength of Lewis acidity. Proton transfer requires Broensted sites that can lose (acid) or accept (base) a proton. The acidity of oxides is related to metal-oxygen bond energy. The general rule is that acidity increases as a function of charge on the metal ion. Acidity of mixed oxides are based on two fundamental assumptions;

1) The coordination numbers of the mixed cations are maintained
2) The coordination number of the anion (oxygen) will be that of the major component oxide.

There will be charge imbalance due to the rare coordination. Acidity will be increased. Bonds can contract due to impurity contamination and X-O-Y bonds exist [325]. The acidity of a surface has a large effect on catalytic behavior of an oxide material.

The air to fuel ratio (λ) window is very important and is generally very narrow. In order to widen this window, cerium oxide is added to the support. Cerium oxide works as an oxygen buffer releasing oxygen under reducing atmosphere thus transforming to Ce_2O_3. Under oxidizing conditions cerium can be re-oxidized to CeO_2.

3.6. Oxides in Three Way Catalysts

Oxide materials used in three way catalysts can be divided to two groups. The first groups being the very stable oxides like silicon dioxide and aluminum oxide that are stable under the reaction conditions. These oxides are using as catalyst supports to disperse catalyst and give it a high surface to weight ratio. The second groups of oxide materials are the transition metal oxides that can catalyze redox reactions by actively taking part in the reactions through transfer of the lattice oxygen of the oxides. These oxides require high oxygen vacancy mobility [326,327]. Nanophase materials have a high vacancy concentration combined with a large surface area and high diffusivity which increases the catalytic activity of the oxide materials.

The spillover phenomenon refers to migration of mobile species across two phases. In three way catalysts, oxygen in particular can be transported between the noble metal islands and the oxide support. Ions can move in both directions depending on the ambient atmosphere. Spillover is strongly dependent on the basicity of the oxide phase as well as the size and morphology of the noble metal particles. For cerium oxide (CeO_2) O spillover is very fast while hydrogen spillover is totally suppressed [328].

Studies of the relation between particle morphology and the structure of the metal support interface, and catalytic properties have been performed on several different metal-support systems for different reactions. Vaarkamp et al. [329] found that high temperature aging of Pt/g-Al_2O_3 changed the Pt particle morphology from three- to two-dimensional, and the particle support interface. Hydrogen in the interface was desorbed. These changes in the structure affected the catalytic, chemisorption and electronic properties. Aging decreased the selectivity of this catalyst for hydrogenolysis of neopentane and methylcyclopentane to methane. At the same time, the activity of the isomerization of neopentane and ring opening of methylcyclopentane increased. On the electronic level, the number of holes in the d-band increased. All these changes in properties are related to the change in structure of the metal support interface.

Chapter 4

4. PREPARATION OF NANOSTRUCTURED MATERIALS

Syntheses of zero dimensional (or so-called quantum dots or nanoparticles), one dimensional (including rods, wires, belts, and tubes), two dimensional and three dimensional nanostructured materials have attracted for the past many years. Because of their size dependent catalytic, electronic, optical, mechanical and optoelectronic properties, these can be broadly tuned through size variation. They play an important role in both interconnects and functional units in fabricating electronic, optoelectronic, electrochemical and electromechanical devices with nanoscale dimensions. The importance of nano-structured materials in catalysis, electrochemistry, functional ceramics, and sensors, their fabrication in nano-size form with anisotropic morphology are considered. When developing a synthetic method for generating nanostructures, the most important issue that one need to address is the simultaneous control over dimensions, morphology (shape) and monodispersity (uniformity) [27,29,296].

Literatures survey indicates that a large number of techniques have been applied for the development of nanostructured materials of varying shape and sizes. Those techniques which are capable to produce fine nanocrystals, are potentially suitable to synthesis nano-structured materials. For consolidation, these methods have been categorized as either physical vapor deposition (PVD), chemical vapor deposition (CVD) or solution-based chemistry (SBC). Each respective method can been subdivided into the individual techniques that are summarized in Scheme as given below in Table 2.

4.1. Physical Vapor Deposition

Physical vapor deposition (PVD) occurs when a material is physically released from a source and transferred to a substrate. The three most important technologies for deposition of nanostructured materials are thermal evaporation, sputtering, and pulsed laser deposition; each of which will be described below.

Table 2. Preparation methods of nanostructured materials

Physical Vapor deposition	Thermal evaporation Rf sputtering method Pulsed laser deposition
Chemical Vapor deposition	Thermal (CVD) Low Pressure (LPCVD) Plasma enhanced(PECVD) Metal-organic(MOCVD) Molecular beam epitaxy (MBE) Atomic layer deposition(ALD)
Solution-based Chemistry	Sol-gel chemical rout Sonochemical method Hydrothermal Synthesis Solvothermal method Homogeneous/heterogeneous precipitation Co-precipitation method Microemulsion method Template-Assisted Synthesis Methods Electrochemical synthesis Electrophoretic deposition (EPD) technique
Miscellaneous Methods	High-energy ball milling

4.1.1. Thermal Evaporation

During thermal evaporation, the substrate, crucible and source materials placed inside a vacuum chamber at room temperature. A vacuum is required to increase the vapor pressure during sublimation and often ranges between 10^{-2} and 10^{-9} Torr (ultra high vacuum). Once the vacuum chamber has stabilized at the appropriate pressure, a heating source is used to heat the source material within the crucible to its vapor point. Upon evaporation, the material will re-deposit along the cooler surfaces of the vacuum chambers, as well as the collection substrate. Typical heating sources include electron-beam, radio-frequency (RF) inductive, and resistive heating.

During electron-beam evaporation, an electron source is aimed at the source material causing localized heating. In comparison, RF induction uses an

AC power supply to produce an alternating current through an induction coil. The alternating current generates a magnetic field within the coil. When the source material is placed inside the coil, the magnetic field induces eddy currents within the source material providing localized heat. Although higher frequencies equate to higher heat rates, lower frequencies are better suited for thicker samples. Finally, resistive heating provides heat by sending a high current source through a resistive coil, such as tungsten, and is a non-localized heat source and therefore commonly used for furnace applications. Although inductive RF and electron-beam sources have been limited to highly-oriented, thin-films of ZnO, resistive sources have produced thin-films, as well as a diversity of ZnO nanostructures with different shapes, sizes, and orientations.

4.1.2. Rf Magnetron Sputtering

Sputtering is the removal of surface atoms via high energy ions. Sputtered films are typically polycrystalline and form at low temperatures with good adhesion properties. Common types of sputtering are focused ion-beam, direct current (DC), radio-frequency (RF), and magnetron. During focused ion-beam sputtering, gallium ions are accelerated through a vacuum towards a sample surface. Acceleration and focusing capabilities are provided by a series of capacitive plates and magnetic coils, respectively. In general, the focused beam of ions provides an exquisite tool for milling and cutting at the nanoscale and has been used within this dissertation as a tool for building prototype nanobelt devices. In comparison, during DC sputtering, the substrate and source (target) material are placed inside a vacuum chamber. Upon evacuation of foreign gases, an inert gas, such as argon, is introduced into the chamber at low pressures. Then, a DC power supply is used to ionize the inert gas in order to produce charged plasma. The ions are accelerated towards the surface of the target, causing atoms of the source material to break off from the target and condense on all surfaces including the substrate. A limitation to DC sputtering is the high voltage required to sputter insulating materials due to the build-up of positive charge on the target material. To solve this problem, the DC power source should be replaced by an RF power source (RF sputtering). In addition, a strong magnetic field (magnetron sputtering) can be used to concentrate the plasma near the target to increase the deposition rate. When applied to ZnO, sputtering has been limited to polycrystalline thin-films [330].

4.1.3. Pulsed Laser Deposition

During pulsed laser deposition (PLD), a laser beam focused through a vacuum onto the surface of a target material. At sufficiently high flux densities and short pulse durations, the target material is rapidly heated to its evaporation temperature and forms a vapor plume. Unlike thermal evaporation, where the vapor composition is dependent on the vapor pressures of the elements within the source material, laser ablation produces a plume of material with similar stoichiometry to that of the target material. Once the vapor plume has been formed, it is collected onto a cooler substrate that promotes nucleation and growth of crystalline films. It is important to note that by using a single crystal substrate, epitaxial single-crystals can be grown that are equal in quality to molecular beam epitaxy. As applied to ZnO, fabrication of (0001) epitaxial films on cubic (111) substrates have been formed using pulsed laser deposition, as well as aligned ZnO nanorod and nanodot arrays without the aid of a catalyst [331-334].

4.2. Chemical Vapor Deposition

During chemical vapor deposition (CVD), the substrate is placed inside a reaction vessel where the pressure and gas flow are controlled. Fundamentally, the process is a chemical reaction between source gases; the product of which condenses during the formation of a solid material within the reaction vessel. The most common CVD techniques used to deposit nanostructured materials are thermal CVD, low pressure CVD (LPCVD), plasma-enhanced CVD (PECVD), metal-organic CVD (MOCVD), molecular beam epitaxy (MBE), and atomic layer deposition (ALD). For each of these methodologies, a vacuum chamber with gas flow control is required. Each methodology is described in detail below.

4.2.1. Thermal and Low Pressure Chemical Vapor Deposition

During thermal and low-pressure chemical vapor deposition (LPCVD), pressures range between 10^{-3} Torr (thermal CVD) and 0.1 Torr (LPCVD) where the reactions occur at excess of 900°C between supplied gases. The processes typically performed simultaneously on both sides of the substrate and produce layers with excellent uniformity and material characteristics.

Although LPCVD is commonly used for depositing thermal oxides within the semiconductor industry, it is limited by its high temperatures and slow deposition rates. However, the method has recently been revitalized through the synthesis of carbon nanotubes , silicon nanowires and well-aligned ZnO nanorods[335-337].

4.2.2. Plasma-Enhanced Chemical Vapor Deposition

In comparison to LPCVD, plasma enhanced CVD (PECVD) is performed at temperatures as low as 300°C. To compensate for the low temperatures within the reactor, plasma is generated to increase the energy available to the chemical reaction. Using PECVD, thick films of ZnO have been prepared [338].

4.2.3. Metal-organic chemical vapor deposition

Metal-organic chemical vapor deposition (MOCVD) is nearly identical to the above listed types of CVD, except a metal-organic precursor is used to catalyze the chemical reaction. The method has been extensively used for the ZnO thin films and aligned nanotips, nanowires and nanotubes [339-342].

4.2.4. Molecular Beam Epitaxy

During molecular beam epitaxy (MBE), pure elements are slowly heated in individual quasi-knudsen cells. Upon evaporation of the individual materials (commonly Gallium and Arsenide), a computer-controlled shutter introduces alternate vapors (beams) into the ultra high vacuum chamber. The term “beam” is used because the mean-free-path of the vapors is sufficiently large to prevent all atomic interaction prior to substrate arrival. Upon condensation onto the single-crystal substrate, atomic layers epitaxiall-grown one layer at a time. During crystal growth, reflective high energy electron diffraction (RHEED) is used to monitor crystal quality and thickness. The exquisite control of MBE has led to the development of structures where the electrons are spatially-confined to quantum wells and quantum dots. Single-crystalline films of ZnO and quantum dots have been grown through this technique [343,344].

4.2.5. Atomic Layer Deposition

Atomic layer deposition (ALD) is a self-limiting, sequential, surface chemistry technique used to deposit conformal thin-films onto substrates of varying composition. Although it is similar to CVD, ALD divides the chemical reaction into two half-reactions by preventing the precursors from interacting within the reaction vessel. For example, when considering alumina (Al_2O_3), the binary reaction in Eq. 5 can be split into Eq. 6 and 7:

$$2Al(CH_3)_3 + 3H_2O -> Al_2O_3 + 6CH_4 \qquad \text{(Eq.5)}$$

$$AlOH^* + Al(CH_3)_3 -> AlOAl(CH_3)_2^* + CH_4 \qquad \text{(Eq.6)}$$

$$AlCH_3^* + H_20 -> AlOH^* + CH_4 \qquad \text{(Eq.7)}$$

where the asterisks denote surface species. From Eq. 6, aluminum is deposited onto a methylated surface by reacting $Al(CH_3)_3$ with hydroxyl (OH^*) species. The reaction is complete either when all of the hydroxyl groups have reacted or when the remaining precursors are evacuated from the reaction vessel. Once aluminum has been deposited, the reaction chamber is purged with an inert gas to prevent mixing of the respective precursors. Then oxygen is deposited using Eq. 7 onto a rehydroxylated surface by reacting H_2O with $AlCH_3^*$. Once again, the reaction is complete either when all of the methyl species have reacted or when the remaining precursors are evacuated from the reaction vessel. By sequentially alternating Eq. 6 and 7, Al_2O_3 will be deposited with atomic layer control [345]. When ALD has been used for ZnO, inverted opals [346] have been grown for photonic applications, as well as thin films [347] for RF MEMS applications.

4.3. Solution-Based Chemistry

Solution-based chemistry (SBC) is defined by any chemical reaction that requires a liquid. As a generic method for synthesis, SBC has been vital to the production of a diversity of materials that are often difficult to make using PVD or CVD. Typically, solution-based methodologies provide materials with high yield and uniformity. However, a common drawback is an increased number of point, line and planar defects when compared to their simpler physical counterparts. The most important and common techniques for nanostructured materials synthesis are co-precipitation, solvothermal, sonochemical, hydrothermal and sol-gel synthesis, which are discussed below.

4.3.1. Sol-Gel Method

The sol-gel process, as the name implies, involves the evolution of inorganic networks through the formation of a colloidal suspension (sol) and gelation of the sol to form a network in a continuous liquid phase (gel). The sol-gel process is a process for making glass/ceramic materials. The sol-gel process involves the transition of a system from a liquid (the colloidal "sol") into a solid (the "gel") phase. The sol-gel process allows the fabrication of materials with a large variety of properties: ultrafine or spherical shaped powders, monolithic ceramics and glasses, ceramic fibers, inorganic membranes, thin film coatings and extremely porous aerogels. Sol-gel chemistry is a remarkably versatile approach for fabricating materials. An overview of the sol-gel process is presented in a simplified chart (Figure 9). The sol is a stable suspension of colloidal solid particles of a diameter of few hundred nm, usually inorganic metal salts, within a liquid phase. For a sol to exist, the solid particles, denser than the surrounding liquid, must be small enough for the forces responsible of dispersion to be greater than those of gravity. Practically, particles in a colloidal sol must have a size comprised between 2 nm and 200 nm; this corresponds to 103 to 109 atoms per particle. In a typical sol-gel process, the precursor is subjected to a series of hydrolysis and polymerization reactions to form a colloidal suspension, then the particles condense in a new phase, the gel, in which a solid macromolecule is immersed in a solvent.

Hydrolysis is a chemical reaction or process, in which a molecule is split into two parts by reacting with a molecule of water. In inorganic chemistry, the word is often applied to solutions of salts and the reactions by which they are converted to new ionic species or to precipitates (oxides, hydroxides, or salts). Many metal ions are strong Lewis acids, and in water they may undergo hydrolysis to form *basic salts*. Such salts contain a hydroxyl group (-OH) that is directly bound to the metal ion in place of a water ligand. At the functional group level, three reactions are generally used to describe the sol-gel process: hydrolysis, alcohol condensation, and water condensation. However, the characteristics and properties of a particular sol-gel inorganic network are related to a number of factors that affect the rate of hydrolysis and condensation reactions, such as, pH, temperature and time of reaction, reagent concentrations, catalyst nature and concentration, H_2O/Si molar ratio (R), aging temperature and time, and drying. Of the factors listed above, pH, nature and concentration of catalyst, H_2O/Si molar ratio (R), and temperature have been identified as most important. When applied to ZnO, nanocrystals, thin films, and nanorods have been synthesized with preferred crystallographic orientations using the sol-gel method [248-253].

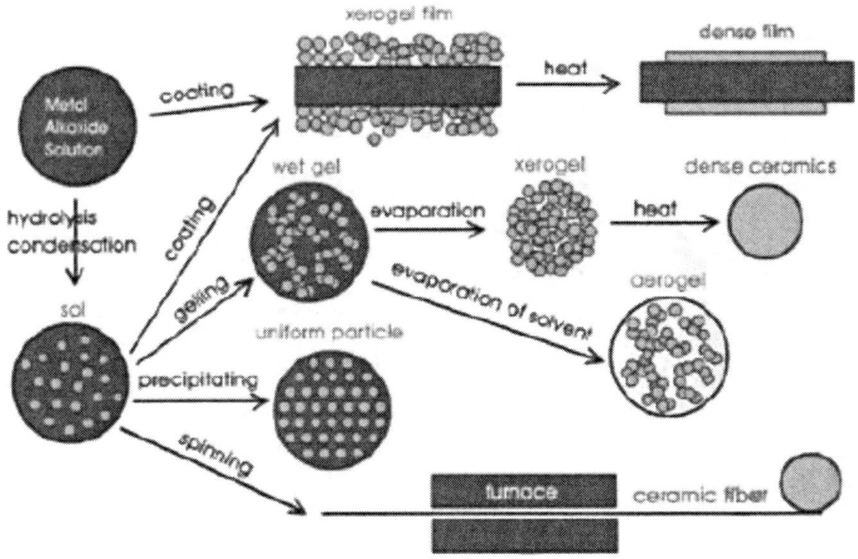

Figure 9. Simplified Chart of sol-gel processes.

4.3.2. Sonochemical Method

Sonochemical method is a cheap method for preparation of nanomaterials for example, in Romania scientist carried out organic reactions using cheap ultrasonic baths as their source of radiation. Sonochemistry allowed the synthesis of a variety of nanomaterials such as metal nanoparticles, metallic colloids, metallic alloys, composite of polymer and metal nanoparticles etc. In Sonochemistry, molecules undergo chemical reaction due to the application of ultrasound radiation (10 KHz-20 KHz). The phenomenon responsible for the sonochemical process is the acoustic cavitation [354]. The main event in the sonochemistry is the creation, growth, and collapse of the bubble that is formed in the liquid. Ag, Pd, Au, Pt and Rh nanoparticles have been synthesized sonochemically in the presence of surfactants like sodium dodecylsulfate (SDS), sodium dodecyl benzene sulfonate (anionic surfactant), polyethylene glycol monostearate, (PEG-MS), and dodecyl trimethyl ammonium bromide [355,356]. The preparation of Pd and Pt nanoparticles by the sonochemical reduction of solutions containing H_2PtCl_3 or K_2PtCl_4 was reported by Fujimoto et al [357]. Colloidal solutions of gold nanoparticles have been prepared from the sonochemical reduction of tetrachloroaurate (III) ions [358]. The reduction rate of auric chloride solution to gold nanoparticles is strongly dependent on the atmosphere, the bulk solution temperature, ultrasound intensity and the reaction vessel-oscillator distance. A comprehensive study on the sonochemical synthesis of colloidal solutions of noble metals is available elsewhere [359-361]. Nanoparticles have been deposited in polymeric materials, microspheres and in amorphous alumina using ultrasound irradiation [362-364]. Some metal oxide nanoparticles have also been prepared using ultrasound irradiation, Zhu and co workers discovered a novel method for preparing highly photoactive nano sized TiO_2 photocatalysts. The method has been developed by hydrolysis of titanium isopropoxide in pure water or a 1:1 $EtOH:H_2O$ solution under ultrasonic irradiation [364,365].

4.3.3. Solvothermal Method

Solvothermal method for synthesis of nanoparticles is almost identical to hydrothermal method but in this no aqueous solvent is used and the temperature can be elevated than in hydrothermal method. Solvothermal synthesis utilizes a solvent under pressures and temperature above its critical point to increase the solubility of a solid and to speed up the reactions between solids. In a typical procedure, a precursor and possibly a reagent (capable of regulating or templating the crystal growth) are added into solvent at the appropriate ratio. This mixture is then placed in an autoclave to allow the reaction and nanowires growth to proceed at elevated temperature and pressure. The solvothermal method normally has better control than hydrothermal methods of the size and shape distributions and crystalinity of nanomaterials. A variety of nanoparticles and nanorods such as CeO_2 and CeOx, V_2O_5, ZnO, TiO_2, SnO_2, In_2O_3, and PbO have been synthesized using solvothermal method with/without the aid of surfactants [366-370].

4.3.4. Hydrothermal Synthesis

As defined, hydrothermal synthesis is a subset of solvothermal synthesis which involves water at elevated conditions. The basic principle is that small crystals will homogeneously nucleate and grow from solution when subjected to high temperatures and pressures. During the nucleation and growth process, water is both a catalyst and occasionally a solid-state phase component. Under the extreme conditions of the synthesis vessel (autoclave or bomb), water often becomes supercritical, thereby increasing the dissolving power, diffusivity, and mass transport of the liquid by reducing its viscosity. In addition, the ability to tune the pressure of the vessel provides an avenue to tailor the density of the final product. When compared to other methodologies, hydrothermal synthesis is environmentally benign, inexpensive and allows for the reduction of free energies for various equilibria.

Materials that are made hydrothermally are generally high-quality, single crystals with a diversity of shapes and sizes. Although hydrothermal synthesis is an established synthesis route within the ceramics industry, it has recently been rekindled within the scientific community by synthesizing one-dimensional nanostructures, such as carbon nanotubes and oxide nanowires. Different types of nanoparticles such as ZnO nanorods, Ag, GaN, TiO_2, $LaCrO_3$, ZrO_2, Sb_2S_3, CrN, α-SnS_2, PbS, Ni_2P and SnS_2 nanotubes, Bi_2S_3 nanorods and SiC nanowires, TiO_2 nanorods have been successfully synthesized in this way [371-375].

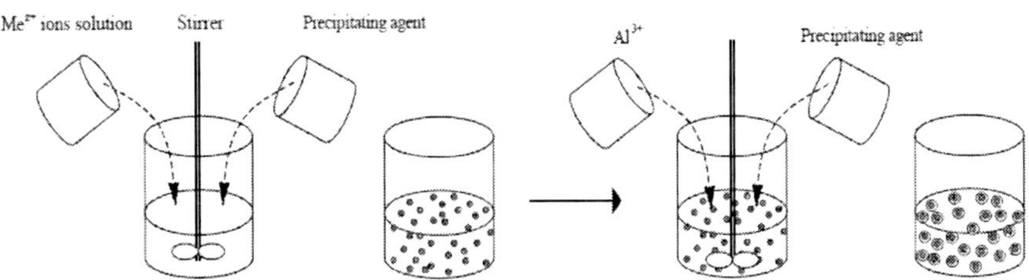

Figure 10. Schematic representation of the sequential precipitation of Al(III) over the already precipitated Ce- Me oxalate particles that serve as seeds.

4.3.5. Homogeneous/Heterogeneous Precipitation

A precipitation phenomenon denotes a relatively rapid formation of a solid product from a solution. Many investigators have dealt with the precipitation category and there are other reference sources for treating precipitation processes. Homogeneous and heterogeneous nucleation refers to the formation of stable nuclei with or without foreign species, respectively. Precipitation refers to more than a nucleation phenomenon. It is not depend on the presence of solute crystalline matter, and in this respect, it does not involve secondary nucleation. Co-precipitation refers to the precipitation in which several constitutes of the solution are found in the same precipitate. It does not result from homogeneous or heterogeneous nucleation process. Precipitation implies a dynamic system, composed of two or more phases. If carried out by rapid mixing of reactant, it is mostly subject to homogeneous nucleation-based phase change. When the catalytic converter operates at high temperatures for a long time, CeO_2-based catalyst is subject to severe temperature conditions, which lead to its particles growth and sintering. Therefore, there is a strong demand for maintaining the thermal stability of the CeO_2-based catalysts. The schematic presentation of this process is presented in Figure 10.

4.3.6. Co-Precipitation Methods for Metal Oxide Synthesis

One of the conventional methods to prepare nanoparticles of metal oxide ceramics is the precipitation method. This process involves dissolving a salt precursor, usually a chloride, oxychloride or nitrate, such as $AlCl_3$ to make Al_2O_3, $Y(NO_3)_3$ to make Y_2O_3, and $ZrOCl_2$ to make ZrO_2. The corresponding metal hydroxides usually form and precipitate in water by adding a base solution such as sodium hydroxide or ammonium hydroxide to the solution. The resulting chloride salts, i.e. NaCl or NH_4Cl, are then washed away and the hydroxide is calcined after filtration and washing to obtain the final oxide powder. This method is useful in preparing composites of different oxides by co-precipitation of the corresponding hydroxides in the same solution. One of the disadvantages of this method is the difficulty to control the particle size and size distribution. Very often, fast (uncontrolled) precipitation takes place resulting in large particles. Nanophase powders of $Y_xZr_{1-x}O_{2-x/2}$ have been prepared from a mixture of commercially available ZrO_2 and Y_2O_3 powders [376]. It was found that depending on the starting powder mixture composition, the yttrium content in the nanophases can be controlled and the tetragonal or cubic phases can be obtained. Tetragonal or a mixture of tetragonal and cubic were observed for low yttria content (3.5 mol% yttria), and cubic for higher yttria contents (19, 54, 76 mol% yttria). These powders were found to have a most probable grain radius of about 10 to 12 nm and the grains appear as isolated unstrained single crystals with polyhedral shapes. The grain shapes appeared to be polyhedral and not very anisotropic. Lattice fringes were parallel to the surfaces demonstrating that (100) and (111) faces dominate.

4.3.7. Micro-Emulsion Method

The term "micro-emulsion" was introduced by Schulman et al., [377] micro-emulsion is made up of water, oil and surfactant. This method has been successfully applied to synthesize metal, alloy, metal suphides and metal oxide nanoparticles. A literature survey depicts that the ultrafine nanoparticles in the size range between 5-50 nm can be easily prepared by this method. This technique uses an inorganic phase in water-in-oil microemulsions which are isotropic liquid media with nanosized water droplets that are dispersed in a continuous oil phase. Synthesis of mixed metal nanoparticles by the micro-emulsion method allows simultaneous control of size and composition. Micro-emulsion is a liquid droplet containing the noble metal precursor being engulfed by surfactant molecules and uniformly dispersed in an immiscibly continuous organic phase which serves as a micro- or nano-scaled reactor in which the chemical reaction takes place. The size of the micro-emulsions is of the order of a few to hundred nanometers and is determined by the balance of the surface free energy mediated by the surfactant molecules and the difference in free energy caused by the immiscibility of the two liquid phases. Reduction of the metal can be carried out by adding the reducing agent directly into the micro-emulsion system, or by the introduction of another reducing-agent containing micro-emulsion. The reducing agent must be stable in an aqueous environment and must not react with the other components of the system. Thus, all non-aqueous reducing agents are excluded. The most commonly employed reducing agents are borohydride and hydrazine. The surfactant molecules serve the role as a protective agent, thus preventing the agglomeration of nanoparticles.

The main attractiveness of the micro-emulsion method is its ease in controlling the size distribution and composition of the metal particles within a narrow distribution by varying the synthetic conditions. The particle sizes depend on the precursor concentration and the amounts of surfactant. Metal alloys can be synthesized if the salts of different metals are dissolved in the solution before the reduction is carried out, provided that the metal salts are miscible in the metallic state. The final composition of the mixed metal nanoparticles can be easily controlled by the ratio of the metal precursor solutions. However, the micro-emulsion method employs costly surfactant molecules and substantial number of separation and washing steps are needed before usage. Thus, this method may not be economical and suitable for large-scale production (Figure 11).

However, one problem associated with this method is the extraction of nanoparticles from the micro-emulsion system. The commonly used method is deposition of the capped nanoparticles on a solid support. An example is the deposition of platinum and palladium nanoparticles synthesized by w/o micro-emulsion on γ-Al_2O_3 alumina support. A common solid support is silica and this support is not directly added but prepared in-situ by hydrolysis and polycondensation of tetraethylorthosilicate (TEOS) with dilute ammonium solution. Deposition of nanoparticles on solid support could have the following two disadvantages. Firstly, the nanoparticles are firmly held by and will not be able to detach from the solid support. Secondly, the reversed micelle is spherical in shape, hence, only nanoparticle but not other 1D nanomaterials can be synthesized. A diverse range of metal nanoparticles have been prepared by this method including Fe, Fe/Au, Pt, Ag, CdS, Pd, Cu, Ni and Au [378-390].

4.3.8. Template-Assisted Synthesis Methods

Among all the synthetic methods for nanomaterials, the template methods are widely used. Most template methods, which confined the growth of materials within a template (such as pores), followed by removal of the template, provide a flexible synthetic route for a variety of nanostructured materials. Both hard-templates (e.g. porous materials, mesoporous silica, track-etched polycarbonate film and carbon nanotubes) and soft-templates such as polymer, self-assembly of surfactant molecules and solvents could be used. Recently, there are few reports on the use of biomolecules such as DNA, tripeptide glutathione (GSH) and lysozyme as template for the synthesis of nanoparticles.

When porous materials are used as templates, the materials should have channels with nanometer scale. Figure 12 illustrates how nanotubes or nanowires can be formed using porous materials template. Firstly, the substrates are loaded into these nano-channels by various methods such as vapor phase sputtering, liquid phase injection, solution-phase deposition or electrochemical deposition. Nucleation, growth and crystallization are confined in these nanometer sized channels. Selective removal of the porous template results in the formation of nanorods, nanowires or nanotubes. Incomplete filling of the channels in the initial step will lead to the formation of nanotubes while complete filling will lead to the formation of nanowires or nanorods. The most commonly used porous materials are porous polymer and alumina films. One typical example is the use of porous anodic alumina (PAA) templates to prepare ZnO nanowires and nanotubes. The zinc-based sol particles were loaded into the PAA nano-channel due to the electrostatic attraction. The ZnO nanowires and nanotubes can be fabricated when the sol particles sintered inside the channels. Although this method is good at obtaining uniform and well-segregated 1D nanomaterials, it is difficult to get nanomaterials with good crystallinity. Outstanding examples of arrays that has been generated by this route are those of oxides nanotubes like TiO_2, In_2O_3, Ga_2O_3, $BaTiO_3$, $PbTiO_3$, In_2O_3 and Fe_2O_3, as well as nanorods of MnO_2, WO_3, Co_3O_4 , V_2O_5, and ZnO. Another widely applicable route to inorganic nanotubes and nanorods is to use CNTs as templates. CNTs have been coated with a thin film of secondary materials that builds up the tube wall of the desire inorganic nanotube followed by removal of the carbon nanotube. Most oxide nanotubes and nanorods, such as V_2O_5, Al_2O_3, WO_3, MoO_2, Sb_2O_3 and MoO_3, ZrO_2, RuO_2, SiO_2, and TiO_2 have been prepared using CNTs as templates [391-401].

Self-assembly of surfactant molecules is another important class of templates. Surfactant molecules can be classified into cationic, anionic, non-ionic and zwitterionic according to their polar head groups. When the concentration of surfactant molecules is below its critical micelle concentration (CMC), the surfactant molecules will only lie up at the liquid-vapor interface or dissolve into the liquid. However, once the concentration is over CMC, these surfactant molecules can self-assemble into spherical micelle, cylindrical micelle or lamellar structure depending on the concentration and types of surfactant molecules. Figure 12 depicts the structure of these micelles. These self-assembled structures can be used as templates for the growth of nanomaterials as shown in Figure 14. Reactants can load into or grow outside the cylindrical template. Subsequently removal of surfactant molecules gives nanorods, nanowires or nanotubes. The uses of surfactant as a template for nanomaterials synthesis have been widely explored. Existing 1D nanomaterials such as nanowires and nanotubes can also be used as a template for the synthesis of other 1D nanomaterials. The substrates can coat onto the nanowire or nanotubes for growth. Removal of the template results in the formation of nanotubes. The substrates can also react with the nanowires so that growth of the new materials is directed by the existing nanowires. For example, single-walled carbon nanotubes (SWNTs) were used as a template for the synthesis of one-dimensional SiC, BN nanostructures, ZnO nanorods, $CoFe_2O_4$ nanowires, β-zeolite nanowires.

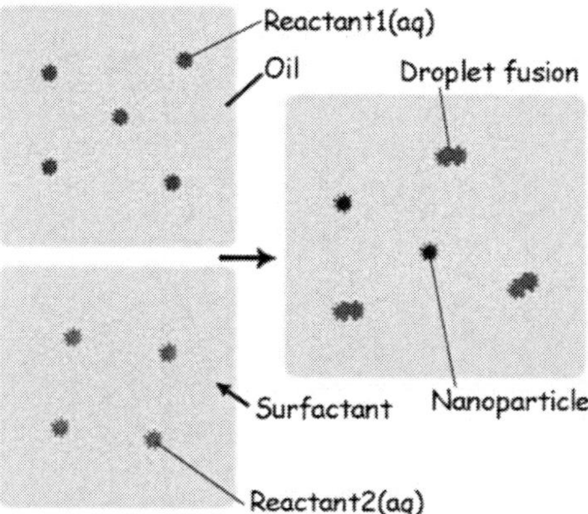

Figure 11. Schematic diagrams on the synthesis of nanoparticles by water-in oil (w/o) microemulsion.

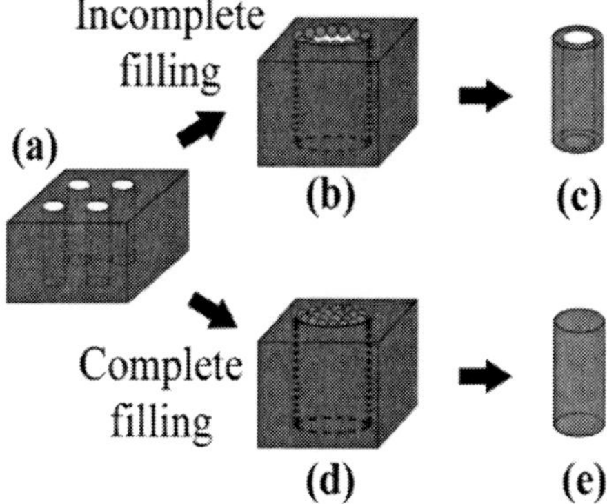

Figure 12. Schematic diagram showing the formation of 1D nanomaterials using porous materials as templates. (a) Porous materials such as porous anodic alumina with nanometer scale channels. (b) Incomplete filling of the substrates in the channel. (c) Formation of nanotube due to incomplete filling. (d) Complete filling of the substrates in the channel. (e) Formation of nanorod or nanowire due to complete filling.

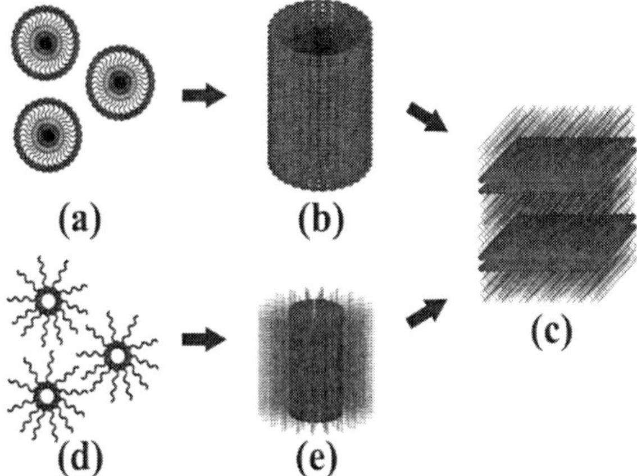

Figure 13. Schematic diagram showing the structures of self-assembled surfactant molecules. Arrows stand for the change of structure as the concentration of surfactant molecules increases. (a) Structure of spherical micelle; (b) Structure of cylindrical micelle; (c) Lamellar structure; (d) Structure of spherical reverse micelle; (e) Structure of cylindrical reverse micelle.

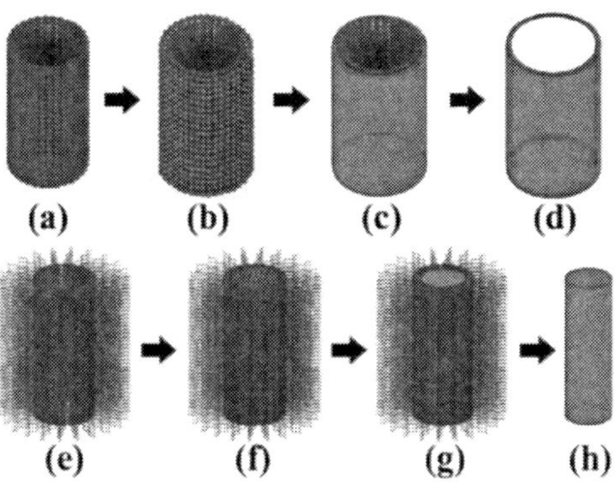

Figure 14. Schematic diagram showing the formation of nanorods, nanowires and nanotubes by cylindrical template. (a) Cylindrical micelle formed by self-assembly of surfactant molecules. (b) Substrate molecules attach onto the outer wall of the cylindrical micelle. (c) Sintering of the substrate molecules form a wall outside the cylindrical template. (d) Selective removal of the surfactant molecules gives the desired nanotubes. (e) Cylindrical reverse micelle formed by self-assembly of surfactant molecules. (f) Substrate molecules load into the cylindrical reverse micelle. (g) Nucleation, growth and crystallization occurs inside the cylindrical template. (h) Selective removal of the surfactant molecules gives the desired nanorods or nanowires.

4.3.9. Electrochemical Synthesis

Electrochemical synthetic technique was established by Reetz and co-workers in 1994 [402-406]. In this method, an anode is made up of the metal whose nanoparticles one interested in and a cathode is of any other metal. Under the suitable applied current density the anode sacrificially dissolves in the electrolyte, the metal ions migrate towards the cathode and reduction occurs. The nucleation and growth of reduced metal atoms occurs at the electrode surface. The stabilizing agent in the reaction vessel arrests the growth of the nanoparticles and facilitates the formation of stable nanostructures. The final step is diffusion of nanoparticles from the electrode surface into the bulk of the solution. Equation 1.1 gives the reactions taking place at the electrodes surfaces.

$$\text{Anode}: M_{bulk} \rightarrow M^{n+} + ne^-$$

$$\text{Cathode}: M^{n+} + ne^- + \text{stabilizer} \rightarrow M_{coll}/_{stabilizer} \qquad \text{(Eq.8)}$$

$$\text{Sum}: M_{bulk} + \text{stabilizer} \rightarrow M_{coll}/_{stabilizer}$$

The advantages of the electrochemical pathway are that the contamination with by-products resulting from chemical reducing agents is avoided and that products are easily isolated from the precipitate. Further, the electrochemical preparation allows for size-selective particle formation. The particle size obtained by electrochemical route depends on many factors, the distance between the electrodes, reaction time, temperature and polarity of the solvent contribute to the particle size. Experiments have also shown that the applied current density also has a major influence on the particle size. Electro-deposition of TiO_2 film from $TiOSO_4$+H_2O_2+HNO_3+KNO_3 (pH 1.4, Eq. 9) solutions involves indirect deposition of a gel of hydrous titanium oxo-hydrides (Eq.11), resulting from the reaction of titanium peroxo-sulfate (Eq.10) with hydroxide ions produced by nitrate electrochemical reduction.

$$NO_3^- + H_2O + 2e^- \rightarrow NO_2^- + 2OH^- \qquad \text{(Eq.9)}$$

$$TiOSO_4 + H_2O_2 \rightarrow Ti\,(O_2)SO_4 + H_2O \qquad \text{(Eq. 10)}$$

$$Ti(O_2)SO_4 + 2OH^- + (x+1)H_2O \rightarrow TiO(OH)_2{\cdot}xH_2O_2 + SO_4^{2-} \qquad \text{(Eq. 11)}$$

A number of investigators have been employed electrochemical method for the development of nanocomposites of Au/polypyrrole, polypyrrole, polyaniline, ZnO, CeO_2, TiO_2, SnO_2 and ZrO_2 [406-410]. The composites system can be prepared by simultaneous reduction of metal ion and autopolymerization of monomers of polypyrrole, polyaniline [411].

4.3.10. Electrophoretic Deposition (EPD)

The phenomenon of electrophoresis has been known since the beginning of the 19th century and it has found application in traditional ceramics technology. EPD is essentially a two-step process: in the first step, charged particles suspended in a liquid migrate towards an electrode under the effect of an electric field (electrophoresis). In the second step, the particles deposit on the electrode forming a relatively dense and homogeneous compact or film. In general, EPD can be applied to any solid that is available in the form of a fine powder (<30 μm) or a colloidal suspension including metals and metal oxides [412].

4.4. Miscellenious Methods (High-Energy Ball Milling)

High-energy milling and mechanical alloying have been studied by several researchers for the synthesis of nanocrystalline materials [413-417]. The basic concept behind this technique is the reduction of the grain size in coarse-grained powder samples to a few nanometers by heavy mechanical deformation followed by powder compaction. Initially the deformation is localized in shear bands with a thickness of about 1μm. These shear bands act as nucleation sites for nanometer-sized grains. However, with increasing milling times, an extremely fine-grain microstructure in the nanometer range with randomly oriented grains separated by high-angle grain boundaries is produced.

Metals with body centered cubic (e.g., Fe, Cr, Nb, W) and hexagonal close-packed (e.g., Zn, Zr, Hf, Co) lattice structures as well as intermetallic compounds (e.g., NiTi, TiAl, Ni_3Al, AlRu) and immiscible systems (e.g., Cu-W) have been milled to nanometer sizes. Gleiter and also Fecht report that high-energy ball milling does not seem to be suitable for metals with face-centered cubic (FCC) crystal structure [269,417]. It was argued that metals with FCC structure are too soft for effective high-energy storage which is a prerequisite to the applicability of this technique. When subjected to high-energy milling, FCC metals were found to sinter to larger particles up to 1mm in size. However, Eckret et al. [416] have been able to synthesis nanocrystalline FCC metals by ball milling in a hydrogen atmosphere. The use of hydrogen during milling seems to embrittle these materials and prevent the problem of softening.

In addition to its suitability to a wide range of metals, alloys, and intermetallics, high-energy ball milling has been reported to be a convenient method to produce large quantities (several kilograms of nanocrystalline materials per day). However, the major drawback of this technique is possibly the high level of impurities arising from the containment vessel and the milling balls.

Chapter 5

5. Characterization of Nanostructured Materials

There are two main categories of nanostructured materials characterizations: structure analysis and property measurements. Structure analysis is carried out using a variety of microscopic and spectroscopic techniques. Characterizations of nanomaterials and nanostructures are largely based on the techniques of surface analysis and conventional characterization methods developed for bulk materials. For example, X-ray diffraction (XRD) has been widely used for the determination of crystallinity, crystal structures and lattice constants of nanoparticles, nanowires and thin films; scanning electron microscopy (SEM) and transmission electron microscopy (TEM, HRTEM) together with electron diffraction are used for the identification of nanoparticles, nanotubes, nanowires, and nanorods. Atomic force microcopy (AFM) which has found wide applications in nanotechnology can produce topographic images of a surface with atomic resolution in three dimensions. The composition of the nanostructures can be examined by energy dispersive X-ray spectroscopy (EDX). X-ray Photoelectron Spectroscopy (XPS)/ X-ray absorption spectroscopy (XAS), selected area electron diffraction (SAED), Reflection Absorption Infrared Spectroscopy (RAIRS), Fourier Transform Raman and Infrared are basic tool to investigate the structural properties of the materials. Photoluminescence (PL)/Fluorescence, UV/Visible absorption spectroscopy are usually employed to investigate the optical properties of nanostructures. Presence of defects can be studied by electron paramagnetic resonance spectroscopy (EPR). *Richard et al.* have edited one encyclopedia to introduce the theory of these characterization techniques in detail

("Encyclopedia of Materials Characterization - Surfaces, Interfaces, Thin Films" edited by C. Richard, A. Jr. Charles, and S. Wilson, Elsevier, 1992).

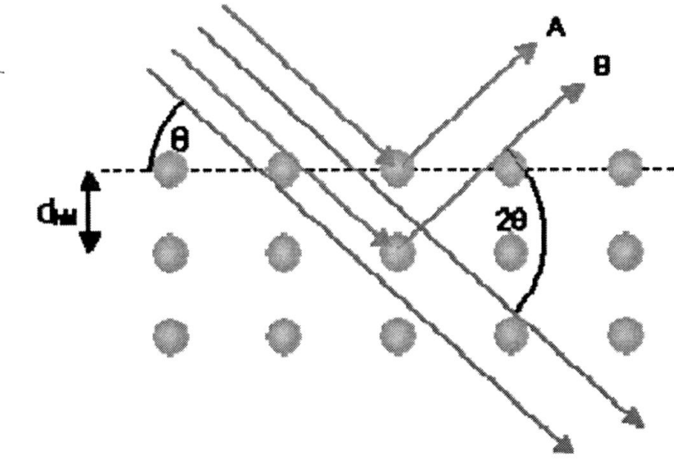

Figure 15. Schematic of diffraction of X-rays crystallography.

5.1. X-Ray Diffraction

X-ray diffraction is one of the most important characterization tool used in solid state chemistry and materials science for investigation crystal structure of solids. X-rays are electromagnetic radiation of wavelength about 1 Å (10-10 m), which is about the same size as an atom. They occur in that portion of the electromagnetic spectrum between gamma-rays and the ultraviolet. The discovery of X-rays in 1895 enabled scientists to probe crystalline structure at the atomic level. X-ray diffraction (XRD) is a noncontact and nondestructive technique which can be used to determine the crystalline phases present in materials, the structural properties of these phases, the thickness of thin films, and atomic arrangements of amorphous materials (Figure 15). Each crystalline solid has its unique characteristic X-ray powder pattern which may be used for its identification. Once the material has been identified, X-ray crystallography may be used to determine its structure, i.e. how the atoms pack together in the crystalline state and what the interatomic distance and angle are etc. X-ray powder diffraction has been widely used to provide information on the following aspects:

(1) The positions and integrated intensities of a set of peaks in an X-ray powder diffraction pattern can be compared to a database of known materials in order to identify the contents of the sample and to determine the presence or absence of any particular phase.
(2) A mixture of two or more crystalline phases can be easily and accurately analyzed in terms of its phase fractions, even if the crystal structures of all phases are unknown. The "so-called" quantitative phase analysis is particularly valuable if some or all of the phases are chemically identical and hence can not be distinguished while in solution.
(3) The crystal structure of unknown material can be determined as a similar material with a known structure exists. Depending on the extent of similarity between new and old structure, this can be fairly straightforward.
(4) The crystal structure of unknown material can be solved ab-initio even if no information is known about the material except its stoichiometry. It is significantly more difficult than the previous case and requires both high-resolution data and efforts to interpret these data.
(5) Phase transitions and solid-state reactions can be investigated in near real time by recording X-ray powder diffraction pattern as a function of time, pressure, and/or temperature.
(6) Subtle structural details such as lattice vacancies of an otherwise known structure can be extracted. Usually it also requires high-resolution data and a very high sample quality.

If X-ray strikes on an atom, it will be weakly scattered in all directions. If it encounters a periodic array of atoms, the waves scattered by each atom will reinforce in certain directions and cancel in others. Geometrically, one may imagine that a crystal is made up of families of lattice planes and that the scattering from a given families of planes will only be strong if the X-ray reflected by each plane arrives at the detector in phase. This leads to the Bragg's law, $n\lambda = 2d \sin\theta$, where λ is the X-ray wavelength, d is the spacing between lattice planes, and θ is the angle of incidence. Note that the angle of deviation of the X-ray is 2θ from its initial direction. This is fairly restrictive for a single crystal: for a given λ, even if the detector is set at the correct 2θ for a given d spacing within the crystal, there will be no diffracted intensity unless the crystal is properly aligned to both the incident beam and the detector. The essence of the powder diffraction technique is to illuminate a large number of crystallites, so that a substantial number of them are in the correct orientation to diffract X rays into the detector.

The geometry of diffraction in a single grain is described through the reciprocal lattice. If the crystal lattices are defined by three vectors a, b, and c, there are three reciprocal lattice vectors defined as:

$$a^* = 2\pi\, b \times c \,/\, a.\, b \times c \qquad \text{(Eq. 12)}$$

and cyclic permutations therefor b* and c* (in the chemistry literature the factor 2π is usually eliminated in this definition). These vectors define the reciprocal lattice, with the significance that any three integers (*hkl*) define a family of lattice planes with spacing $d = 2\pi / \mid h\mathbf{a}^* + k\mathbf{b}^* + l\mathbf{c}^* \mid$, so that the diffraction vector $\mathbf{K} = h\mathbf{a}^* + k\mathbf{b}^* + l\mathbf{c}^*$ satisfies $|\mathbf{K}| = 2\pi / d = 4\pi\sin\theta / \lambda$ (most chemists and some physicists define $|\mathbf{K}| = 1 / d = 2\sin\theta / \lambda$).

The intensity of the diffracted beam is governed by the unit cell structure factor, defined as:

$$\mathbf{F}_{hkl} = \sum_j e^{i\mathbf{K}\cdot\mathbf{R}j} f_j\, e^{-2W} \qquad (2)$$

where R_j is the position of the jth atom in the unit cell, the summation taken over all atoms in the unit cell, and f_j is the atomic scattering factor, tabulated in, e.g. the international Tables for crystallography and is equal to the number of atomic electrons at $2\theta = 0$, decreasing as a smooth function of $\sin\theta/\lambda$. The Debye-Waller factor $2\mathbf{W}$ is given by $2\mathbf{W} = K^2 u^2_{rms}/3$, where u_{rms} is the three-dimensional root-mean-square deviation of the atom from its lattice position due to thermal and zero-point fluctuations. Experimental results are often quoted as the thermal parameter **B**, defined as $8\pi^2 u^2_{rms}/3$, so that the Debye-Waller factor is given by $2\mathbf{W} = 2\mathbf{B}\sin^2\theta/\lambda^2$. The Debye equation shows that the diffracted intensity depends on the distribution of inter-atomic distance within the scattering volume. While the internal structure can be used to calculate a set of inter-atomic distances, the converse may be not necessarily true. If a model diffraction pattern agrees well with observed data, then there is strong evidence that the structure represented by the model stands for the actual structure of the sample.

5.2. X-Ray Photoelectron Spectroscopy

To rationalize the catalytic properties of nanoparticles, the surface composition and structure of the nanoparticles are required. Quantitative XPS analysis is a powerful tool in interpretation of the surface composition. From the spectra of the quantitative XPS analysis of bimetallic nanoparticles, the amount of different metals presence in surface region can be calculated. The photoelectric effect leads to the production of photoelectrons that are emitted if the energy of absorbed photons is high enough. Excess energy is transferred to the electron as kinetic energy. The binding energy is determined by the difference between energy of the incident photon and kinetic energy of emitted electron. High energy X-rays are required to study the properties of core electrons of the atom. X-ray photoelectron spectroscopy (XPS) consists of a source of fixed energy radiation to emit X-rays. A monochromatic source must be used to get high energy X-rays. The source should have multiple-anode handling capacity and high lifetime. An electron energy analyzer is used to disperse the emitted electrons based on their kinetic energy and measure the flux of emitted electrons of a particular energy. It requires precision energy measurements over a wide range of bond energies of elements and ability to define the analysis area. Clean vacuum pumps operating at a base pressure 10^{-10} torr (10^{-8} Pa) are required to remove the adsorbed gases from the sample, to increase the mean free path for electrons analyzer, to improve the efficiency of the X-ray and electron optics.

5.3. X-Ray Absorption Spectroscopy (XAS)

X-ray absorption spectroscopy (XAS) is a family of spectroscopy techniques for determination oxidation states of metals in complex materials. It consists of two spectroscopy techniques: the x-ray absorption near edge structure (XANES) –or NEXAFS alternatively and extended x-ray absorption fine structure (EXAFS) -or SEXAFS alternatively. The XANES and EXAFS involve in measurements of the X-ray photoabsorption of a selected element as a function of energy below and above its core-shell electron binding energy, respectively. Therefore, the XAS provides structural information such as chemical bonding information, bond distances, coordination numbers, local structure of bulk solids, even with powder samples-crystalline or amorphous.

For X-ray energies greater than the binding energy, the absorption process leads to the excitation of the core electron to the ionization continuum. The resulting photoelectron wave propagates from the X-ray absorbing atom and is scattered by the neighboring atoms. The EXAFS spectrum results from the constructive and destructive interference between the outgoing and incoming photoelectron waves at the absorbing atom. The interference gives rise to the modulatory structure –i.e., peaks and valleys of the X-ray absorption versus incident X-ray energy. This process also makes EXAFS unique –i.e, the absorbing atom acts as both the source and detector of the interference; that is EXAFS phenomenon.

5.4. Electron Paramagnetic Resonance Spectroscopy

Electron paramagnetic resonance (EPR) or electron spin resonance (ESR) spectroscopy is a technique to study paramagnetic species that have one or more unpaired electrons, such as free radicals, free electrons, and defects in materials. Electron has magnetic moment and spin quantum number s = 1/2, with magnetic components ms = +1/2 and ms = -1/2. In the presence of an external magnetic field, the magnetic moment of unpaired electrons will align themselves either parallel (ms = -1/2) or antiparallel (ms = +1/2) to the field, so two distinct electron spin states will be formed. The energy difference between them (ΔE) is directly proportional to the magnetic field's strength (B), which is governed by Zeeman interaction: $\Delta E = h\nu = g\beta B$ (1-1) where h is Planck constant, ν is microwave frequency, g is Landé g-factor (which is around 2), and β is Bohr magnetron constant. In EPR measurement, microwave at a fixed frequency is applied to the specimen and the applied field strength is varied. Increasing the external magnetic field will widen the gap between those two electron spin states. When the gap is equivalent to the energy of the microwaves, the electrons in the lower energy state will be excited to the higher energy state and microwave radiation will be absorbed during the transition. Finally, the EPR spectrum can be obtained and the g-factor can provide information on the properties of paramagnetic species in the specimen.

5.5. Energy Dispersive X-Ray Spectroscopy

It is often necessary to identify the different elements associated with a specimen. SEM combined with a built-in spectrometer called an Energy Dispersive X-ray Spectrometer (EDX) is used for this purpose. EDS can provide rapid qualitative or with adequate standards quantitative analysis of elemental composition within a sampling depth of 1-2 microns. EDX can also quantify the elements it detects by a standard or standardless analysis: in standard quantification, the peak intensity of each element in the sample is compared to the peak intensity obtained from a reference standard measuring under same conditions; in standardless quantification, the elemental information can be calculated. In EDX, the electron beam incident on the specimen creates secondary electrons and leaves thousands of holes in the electron shells. Electrons from outer shells will drop into the inner shells and lose some energy in the form of X-rays at the same time. The X-rays emitted from the sample atoms are characteristic in energy and wavelength to, not only the element of the parent atom, but which shells lost electrons and which shells replaced them. The elemental composition of a sample can be identified by a nondestructive technique. These X-rays can be detected by an additional detector attached to the SEM. The analysis of the chemical composition of small details found in SEM images by EDS is extremely useful in most metallographic and biomaterial studies.

5.6. Electron Microscopy

Electron microscopic observations can provide a straightforward evaluation on pore size, shape and distribution in the nanoporous materials. Generally, electron microscopes use a beam of highly energetic electrons to examine objects on a very fine scale. This examination can yield the following information:

- *Topography*: The surface features of an object or "how it looks", its texture;
- *Morphology*: The shape and size of the particles making up the object;
- *Composition*: The elements and compounds that the object is composed of and the relative amounts of them;

- *Crystallographic Information*: How the atoms are arranged in the object;

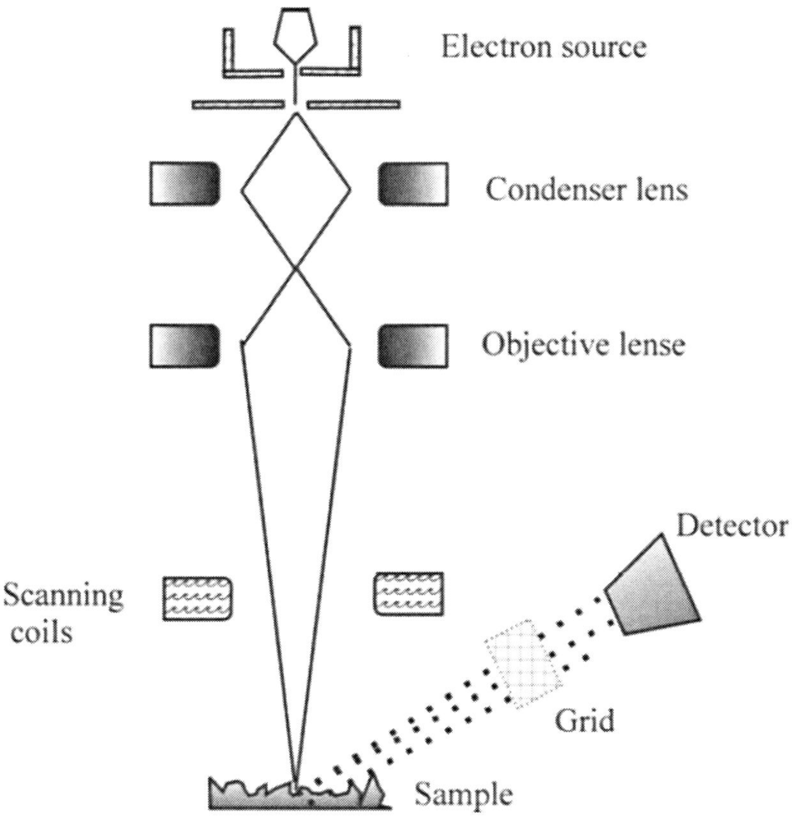

Figure 16. Schematic set-up of scanning electron microscope.

There are two types of electron microscopy: The transmission electron microscope (TEM) was the first type of electron microscope to be developed and is patterned exactly on the light transmission microscope except that a focused beam of electrons is used instead of light to "see through" the specimen; the first scanning electron microscope (SEM) debuted in 1942. Its late development was due to the electronics involved in "scanning" the beam of electrons across the sample. For mesoporous materials, TEM is the most useful method to provide direct observation of pore structures and parameters.

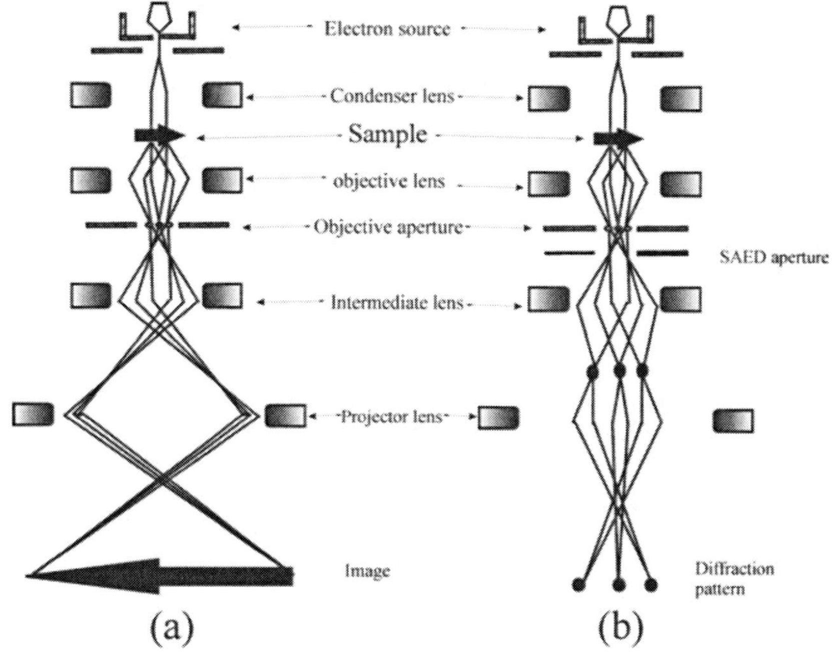

Figure 17. Schematic diagram of a transmission electron microscope: (a) TEM imaging set-up,(b) selected area diffraction pattern processing set-up.

5.6.1. Scanning Electron Microscopy (SEM)

Topographic details of nanostructured materials and film surface can be revealed with great clarity and detail by utilizing the SEM. The SEM is one of the most frequently used instruments in materials research today because of the combination of higher magnification, larger depth of focus, greater resolution and ease of sample observation. Morphology details of less than 50nm can be resolved by SEM and it possesses a depth of focus more than 500 times higher than that of the optical microscope at equivalent magnification. In SEM, electron beam, accelerated by a relatively low voltage of 1-20 kV, is scanned on the specimen surface. As the electron beam strikes the surface, a large number of signals are generated from (or through) the surface in the form of electrons or photons. These signals emitted from the specimen are collected by detectors to form images and the images are displayed on a cathode ray tube screen. There are three types of images produced in SEM: secondary electron images, backscattered electron images, and elemental X-ray maps. Secondary electrons (SE) are considered to be the electrons resulted from inelastic scattering with atomic electrons and with the energy less than 50 eV; Backscattered electrons (BSE) are considered to be the electrons resulted from elastic scattering with the atomic nucleus and with the energy greater than 50 eV. The backscattering will likely occur in a material of higher atomic number, so the contrast caused by elemental differences can be built up. After the primary electron beam collides with an atom in the specimen and ejects a core electron from the atom, the excited atom then decays to its ground state and emit either a characteristic X-ray photon or an Auger electron. The electron beam used for SEM possesses lower energy as compared to TEM. SEM usually has resolution of 1 nm. Typical acceleration potentials used are between 1-50kV. Using a set-up of a number of lenses, the electron beam is focused on to the sample in a single spot of cross section 2-10 nm. The energy dispersive X-ray detector (EDX) can sort the X-ray signal by energy and produce elemental images, so the spatial distribution of particular elements can be detected by SEM (Figure 16).]

5.6.2. Transmission Electron Microscopy (TEM)

Transmission electron microscopy (TEM) is the premier tool for understanding the internal microstructure of materials at the nanometer level. Although X-ray diffraction technique generally provides more quantitative information than electron diffraction techniques, the latter has an important advantage over X-ray in that it can be focused using electromagnetic lenses. This allows one to obtain real-space images of materials with resolution on the order of a few tenths to a few nanometers, depending on the imaging conditions and simultaneously to obtain diffraction information from specific regions in the images as small as 1 nm. Variations in the intensity of electron scattering across a thin specimen can be used to image strain fields, defects such as dislocations and second-phase particles and even atomic columns in materials under certain imaging conditions. In TEM, electron beam emitted from the electron gun accelerated by accelerating voltage of 200 kV. The focused beam is incident on the specimen surface and interacts with the specimen (Figure 17). The transmitted electrons form an image and the final enlarged image is projected on a fluorescent screen. There are two observation methods offered by TEM: diffraction mode and image mode. The diffraction pattern is formed at the back focus plane of the objective lens and the image is formed at the image plane of the objective lens. In diffraction mode, the electron diffraction patterns are different for different materials: the diffraction pattern of a single crystal is a spot pattern; the diffraction pattern of a polycrystal is a ring pattern; and the diffraction pattern of a glassy or amorphous material is a series of diffuse halos. In image mode, the contrast in image can result from several mechanisms: mass contrast from separation of different components; thickness contrast from non-uniform thickness; diffraction contrast from scattering of incident electron wave by structural; and phase contrast in the form of periodic fringes. Selected area electron diffraction (SAED) pattern can be obtained by TEM when a selected-area aperture is used to limit the diffracting volume and the condenser lens is defocused to produce parallel illumination at a specimen. SAED pattern is a superposition of diffraction patterns from crystallites possessing distinct orientations and can be used to determine the Bragg lattice and lattice parameters of crystalline materials. The relationship between the diffracted scattering angle θ measured in an SAED pattern, the associated atomic inter-planar spacing d and the wavelength can be determined using Bragg's Law, $n\lambda = 2d\sin\theta$.

5.7. Scanning Probe Microscopy (SPM)

SPM is unique among imaging techniques which provides three-dimensional (3D) real-space images and among other analysis techniques which allows spatial localized measurements of structure and properties, under optimum conditions. SPM is a general term for a family of microscopes depending on the probing forces used. Now many types of scanning probe microscope, such as scanning confocal microscope and scanning acoustic microscope, have been developed. Not only is the scanning probe microscope a powerful imaging tool with atomic resolution, it is also widely used to investigate the optical, mechanical, electrical, magnetic and piezoelectric property of various materials. Two other major members, scanning tunnel microscopy (STM) and atomic force microcopy (AFM). AFM is a surface analytical technique that can generate nanoscale topography by scanning a fine silicon tip across the surface. Because it can perform high-resolution imaging under physiological conditions in a nondestructive manner, AFM has developed into a powerful tool for studying the structural details of biological systems. Since its invention, AFM has proved to be extremely useful in characterizing various kinds of surfaces on the micrometer to the atomic scale being most powerful in the nanoscale. In addition, AFM can also be used as a nanoscale robot, i.e. for modifying surfaces or manipulating structures such as nanoparticles on surfaces.

The mechanical properties of nanostructures have been successfully investigated by the nano-indentation technique with SPM. In this technique, SPM can monitor the repulsive force between the sharp probe and material surface as a means to detect the mechanical properties of materials. The key advantages of SPM nano-indentation are its imaging capability which allows accurate measurement of indentation geometry and precise location of indentation probe for nanomechanical measurement. This particular study explores the areas of applicability of SPM for measuring mechanical properties such as Young's modulus of nanomaterials.

5.8. Raman / Infrared Spectroscopy

Identification of specific types of chemical bonds or functional groups based on their unique absorption signatures is possible by Fourier Transform Raman/Infrared spectroscopy (FTRaman/FTIR). Infrared absorbance of common functional groups shows the structural behavior of the materials. In nanosize powders the surface species absorption can compete with the absorption of the bulk species. Therefore, the characterization of these powders cannot be complete without the identification of the surface chemical species. FTIR can be utilized to analyze the surface contamination. Typically there will be contamination from atmospheric CO_2, water and hydrocarbons. Especially oxide surfaces that are always more or less hydrolyzed can easily adsorb water molecules and CO_2 to form carbonate species. For example, metal oxides produce infrared bands at around 500-300 cm^{-1}due to O-M stretching. The energy range for infrared spectra measurements is 400-4000 cm^{-1} and the resolution 4 cm^{-1}. Samples mixed with potassium bromide and then ground and pressed to pellets.

5.9. UV-Vis Spectroscopy

UV/visible spectroscopy is the finest technique for characterization of semiconductor nanostructured materials (quantum dots and metal oxides), gold and silver nanoparticles which are absorb radiation in the visible region.

Absorption of ultraviolet and visible radiation by molecules generally occurs in one or more electronic absorption bands, each of which is made up of numerous closely packed but discrete lines. Each line arises from the transition of an electron from the ground state to one of the many vibrational and rotational energy states associated with each excited electronic energy state. Some inorganic materials exhibit charge-transfer absorption. In most charge-transfer materials involving a metal ion, the metal serves as the electron acceptor, such as O^{2-}*(2p)* transferring electron to Ce^{4+}(4*f* orbital)in CeO_2 observed band at around 300 nm.

Enormous numbers of inorganic, organic and biochemical species absorb ultraviolet or visible radiation and are thus amenable to direct quantitative or qualitative determination. It has a high sensitivity with a typical detection limit from 10^{-4} to 10^{-6} M and can be used to determine the concentration of biological macromolecules in solution. It can also be used to monitor protein stability and to follow structural changes such as unfolding and refolding under a variety of conditions. By using rapid mixing techniques (stopped-flow), reactions that occur in the millisecond range can be monitored by UV-vis. Moreover, this method is particularly important for those extremely expensive samples such as proteins and enzymes since the measurements can be carried out in solution, and only tiny amounts of samples are required for analysis. The method is non-destructive and the samples can be recovered afterwards.

5.10. PHOTOLUMINESCENCE

Photoluminescence (PL) is a commonly used characterization technique for materials analysis. PL can be used to determine the composition of a material qualitatively or semiquantitatively, even trace chemicals, impurities, and defects can be identified by PL; PL can also provide information on the structure and electronic states of the materials. In PL, absorption of light results in a transition from the ground state to an excited state of an atom or molecule; then the system undergoes a nonradiative internal relaxation and excited electron moves to a more stable excited level; after a characteristic lifetime in the excited state, the electronic system will return to the ground state. The energy is released in the form of light, and this emitted light is detected as photoluminescence. In the PL spectrum, the spectral dependence of its intensity can provide information about the properties of the material; the time dependence of the emission can provide information about energy level coupling and lifetimes. The presence of defects in the materials can also be estimated by the PL emission involving defect level transitions, so the quality of materials can be determined.

5.11. Fluorescent Microscopy

Fluorescence is a rapid process of emission of light of longer wavelength from a material when it absorbs a light of given wavelength. The fluorescence process occurs in certain molecules called fluorophores or fluorescent dyes. Different fluorescent probes can be attached to the biological specimens who can emit light when they absorb a specific wavelength of light. This procedure is extremely rapid and takes only 10-12 seconds. Different fluorescent molecule possesses different characteristic absorption and emission spectra. A fluorescent microscope consists of a light source (usually a mercury lamp), a filter which enables only a certain wavelength to pass, a microscope objective to project a diffraction-limited image at a fixed plane, and a dichroic mirror which reflects light shorter than a certain wavelength, and passes light longer than that wavelength and a emission filter used to select the emission wavelength of the light emitted from the sample and to remove traces of excitation light.

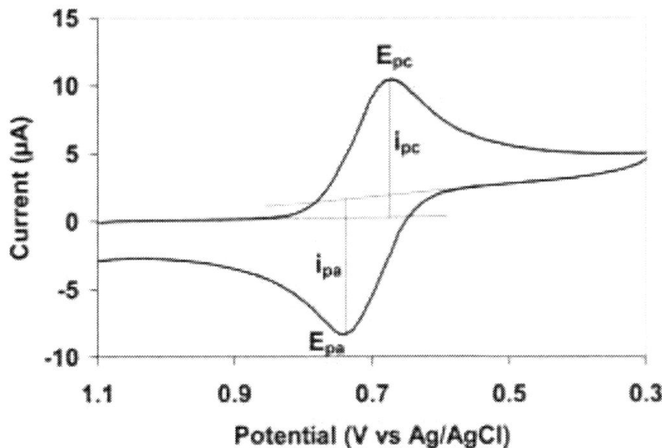

Figure 18. Schematic diagram of cyclic voltammogram.

5.12. Cyclic Voltammetry

Cyclic voltammetry is a kind of potentio-dynamic electrochemical measurement. Being a specific type of voltammetry, it is used for studying the redox properties of the materials like as TiO_2, ZnO, CeO_2 and SnO_2 [418]. For the majority of experiments the electroactive species is in the form of a solution. The three-electrode method is the most widely used because the electrical potential of reference does not change easily during the measurement. This method uses a reference electrode, working electrode, and counter electrode (also called the secondary or auxiliary electrode). Electrolyte is usually added to the test solution to ensure sufficient conductivity. The combination of the solvent, electrolyte and specific working electrode material determines the range of the potential. In cyclic voltammetry, the electrode potential follows a linearly ramping potential vs. time. Measured the potential between the reference electrode and the working electrode and measured the current between the working electrode and the counter electrode. This data, then plotted as current (i) vs. potential (E). The forward scan produces a current peak for any analytes that can be reduced through the range of the potential scan. The current will increase as the potential reaches the reduction potential of the analyte, but then falls off as the concentration of the analyte is depleted close to the electrode surface. As the applied potential is reversed, it will reach a potential that will reoxidize the product formed in the first reduction reaction, and produce a current of reverse polarity from the forward scan. This oxidation peak will usually have a similar shape to the reduction peak. As a result, information about the redox potential and electrochemical reaction rates of the compounds is obtained (Figure 18).

5.13. Electrochemical Impedance Spectroscopy (EIS)

Small-amplitude ($V<kT/q$) ac impedance measurements are used to determine the bulk, and both the electrode and the grain boundary characteristics of materials [419,420]. The grain boundaries respond to a signal applied perpendicular to the boundary, as a capacitor and a resistor in parallel. For electrodes, the charge transfer process under small ac signals also exhibits a response that can be described by a capacitor connected with a resistor in parallel. To assist in identifying the phenomena which determine the ac

impedance spectrum one can vary parameters that affect the relaxation processes differently. The typical parameters are temperature, oxygen partial pressure, dc voltage applied on top of the ac one, length of mixed ionic and electronic conductor (MIEC) and grain size. Raising the temperature will enhance the contribution of the process with the higher activation energy. Lowering the oxygen partial pressure, will promote oxygen diffusion limitation and the corresponding impedance. Applying a dc voltage does not affect the impedance of components which are ohmic, but changes the impedance of the nonohmic components. In particular, when the current depends exponentially on the voltage, as in the Butler-Volmer relations, the dc bias reduces the ac impedance. For a diffusion-limited process the dc bias increases the ac impedance. Increasing the grain size reduces the total grain boundary surface area per unit volume and should therefore reduce the impedance with the grain boundaries. The shape of the ac impedance plots may deviate from that expected for the simple RC and Warburg elements. There are different reasons for deviations. Typical reasons are rough surfaces, constriction resistance, and distribution of elements with different characteristic parameters, mainly in the bulk. The constriction resistance is due to a smaller contact area of the electrode than the nominal electrode area. It has also been shown that for a MIEC electrode the impedance of transfer of oxygen from the gas phase into the MIEC and the impedance of diffusion inside the MIEC, though coupled in series, do not yield separate parts in the Nyquist plot (imaginary part of impedance vs. real part of impedance).

Chapter 6

6. NANO-STRUCTURED MATERIALS FOR BIOMEDICAL APPLICATIONS

Remarkable optical, electrical and mechanical properties of nanostrctured materials provided a new set of materials, which recently have grabbed biologists and biomedical engineers' attention for applications such as biolabels, biosensors and image contrast agents [2-28,421,422]. These nanostructured materials have also been used to fabricate devices for applications such as drug delivery and medical therapeutics. A suitable control of properties at nanoscale structures can lead to new understanding, devices, and technologies. The development of nanoscale devices provides dense information storage, ultra small sensing elements, and data processing elements with reduced dimensions. The large surface area and shape controlled properties of nanostructured materials make them suitable candidates for biomedical device development [421-426]. Therefore, the diversity in shape composition of the nanomaterials (particles, rods, wires, tubes, cubes, tetrapods, or triangles), and the readiness for surface functionalization (physical, chemical, or biological) has enabled the fabrication of various functional nanoscale devices. Biologists have recently begun to borrow these nanotools and apply them to a variety of applications ranging from diagnosis of disease to gene therapies. Integration of biomaterials (e.g. proteins, peptides, or DNA) with semiconductor quantum dots (QDs) and metal nanoparticles (NPs) greatly expands the impact of biophotonics and bioelectronics, particularly in optical imaging and biosensing, as well as therapeutic strategies.

For example, gold is considered to be a noble metal in bulk state, but the nanoparticles of gold are considered as good catalyst. Nanoparticles of gold, silver and semiconductors (quantum dots) such as CdSe, CdS etc can be used in diagnosis of DNA, proteins and other biomolecules. Some nanostructured fluorescent materials such as LnF_3:Ln, $NaYF_4$:Ln, $LnPO_4$ and their functionalized doped materials have been utilized for protein labeling and quantitative DNA hybridization detection. Due to strong superparamagnetic behavior of iron oxide nanoparticles have been employed as a targeting material in biomedical sciences. Immunoglobulin G (IgG) and streptavidin was conjugated to CdSe QDs with different emission spectra to label the breast cancer marker Her2 on the surface of live cancer cells. They also used the conjugated nanoparticles to stain action and microtubule fibers in the cytoplasm, as well as to detect nuclear antigen inside the nucleus. Here, we introduce some of the latest examples of nanostructures that have been applied successfully to problems in biotechnology, with a special focus on biosensing, bioimaging and drug delivery.

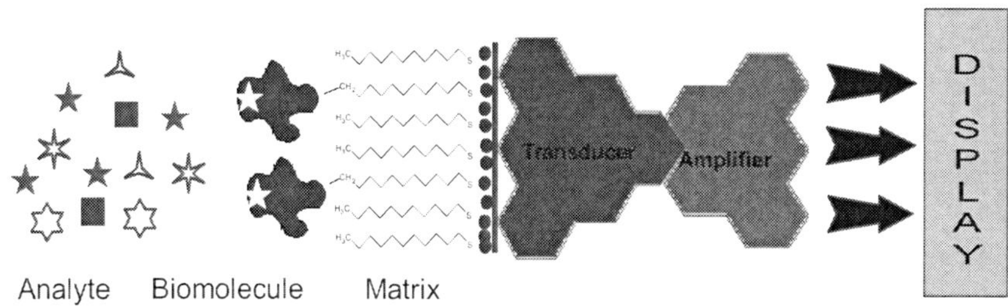

Figure 19. Schematic of nanomaterials based biosensor.

6.1. Biosensors

Biosensors are analytical devices that work on two fundamental operating principles: biological recognition and sensing. Thus biosensor detects a molecular recognition and transforms it into a detectable signal using a transducer e.g. optical, electrochemical, piezoelectric and gravimetric transducers are used to get signals in these devices. So the central theme of detection is the signal transduction associated with the selective recognition of a biological or chemical species. There are two different types of biosensors: biocatalytic and bioaffinity-based biosensors.

Table 3. Various types of nanostructured materials and their characteristics

Nanomaterials	Properties	Functions	Sensors improvement
Metallic nanoparticles (Au, Ag, Pd and Pt etc)	Surface plasmon resonance (SPR), catalytic, scattering, surfaces enhanced fluorescence	Biomolecule immobilization and catalysis, enhancement of electron transfer, labelling of biomolecules	Improved sensitivity, stability and direct electrochemistry of proteins
Fluorescent nanoparticles (LnF_3, LnF_3:Ln, $LnPO_4$, $LnPO_4$:Ln, $NaLnF_4$, $NaLnF_4$:Ln, (Ln = La-Lu), YF_3, YF_3:Ln, $NaYF_4$, $NaYF_4$:Ln)	Strong fluorescent signature due to narrow emission spectra and highly photostability	Biolabeling, biomarkers, bioimaging	Improved photastability, high sensitivity and low detection limit
Semiconductor quantum dots (CdS, CdSe, CdTe, GaN, ZnTe, ZnS, ZnSe, PbS, PbSe, CuS and silica etc.)	Quantum confinement tunable fluorescence signatures, narrow emission spectra, brighter emission and good photostability	Labeling of biomolecules, indirect detection	Improved sensitivity and selectivity
Metal oxide nanoparticles (Al_2O_3, CdO,CeO_2, Cu_2O, Eu_2O_3, Fe_3O_4, Fe_2O_3, FeO, In_2O_3, Mn_3O_4, MnO_2, Pr_2O_3, SiO_2, SnO_2, Sb_2O_3, TiO_2,WO_3, ZnO and ZrO_2)	Magnetic and electrocatalytic property, biocompatibility, chemical reactivity of surface sites	Easy separation, enhance the reusability	Improved sensitivity and selectivity
Organic-inorganic nanocomposites	Catalytic property of inorganic nanoparticles, biocompatibility, hydrophobicity and high mechanical stability	Biomolecule immobilization and catalysis	Stability of bioelectrodes, Improved sensitivity
Polymeric Materials	High electron transport property, biocompatibility, good chemical, mechanical and electrical stability	Drug delivery, biosensors	Better sensing characteristic highly sensitive and rapid electrochemical resonse
Carbon nanotubes (SWCNT,MWCNT)	High surface to volume ratio, fast electron transport behavior	Enhancement of electron transfer from biomolecule to electrode surface	Improved sensitivity and selectivity

The biocatalytic biosensor uses mainly enzymes as the biological compound, catalyzing a signaling biochemical reaction. The bioaffinity-based biosensor, designed to monitor the binding event itself, uses specific binding proteins, lectins, receptors, nucleic acids, membranes, whole cells, antibodies or antibody-related substances for biomolecular recognition. Many kinds of nano-structured materials including polymers, metal, metal oxides, semiconductors and organic-inorganic nano-structured composite materials have been utilized for fabrication of biosensor. Based on their unique physical, chemical and electrocatalytic properties, these nanoparticles play variety of roles in different biosensing systems. Some specific properties of nanostructured materials such as high surface area, biocompatibility, non-toxicity, electrocatalytic and piezoelectricity that allow transducers using either bulk acoustic waves (BAW) or surface acoustic waves (SAW) to measure changes in fundamental frequencies. Owing to their excellent film forming and adhesion ability, high surface area, strong adsorption capability, improved catalytic efficiency (oxygen storage capacity), better chemical stability, small grain size, which make them highly promising for biosensor applications. The basic functions of nanoparticles can be mainly classified as immobilization of biomolecules, catalysis of electrochemical reactions, enhancement of electron transfer, labelling of biomolecules and acting as reactant.

The biorecognition element determines the degree of selectivity or specificity of the biosensor, whereas the sensitivity of the biosensor is greatly influenced by the transducer (scheme 19). According to the transduction mechanism of nanomaterials have been classified and summarized into following sections (Table 3).

6.1.1. Colorimetric Sensors

These devices are mainly based on the observation of color changes due to anion-cation interactions and for monitoring colorimetric reactions and for metal detection. Fundamentally, molecular recognition involves the interactions between molecules: i.e. bond formation, acid–base interactions, hydrogen-bonding, dipolar and multipolar interactions, π–π molecular complexation and last and least, vander Waals interaction and physical adsorption. The oligonucleotide-mediated nanoparticle aggregation process has been extensively used for the development of simple and highly sensitive colorimetric biosensors for oligonucleotides by Mirkin and others [426-434].

The detection of specific oligonucleotide sequences is now very important in diagnosis of genetic and pathogenic diseases and quantifying the amount of product generated by polymerase chain reaction (PCR). The general procedure for detection of oligonucleotides is through the fabrication of nanoparticles, functionalized with single-stranded DNA. Upon addition of the target sequence the particles aggregate, changing the color of the solution. Using this method oligonucleotides were detected at sub-picomolar level without the assistance of PCR. This methodology was also applied for the colorimetric screening of DNA binders and triplex DNA binders. Based on the similar approach a highly selective and sensitive lead (Pb^{2+}) biosensor was reported by Liu et al. [433-436]. In their sensor design, they used a Pb^{2+} specific ''DNAzyme''composed of a catalytic and a substrate strand. In the presence of Pb^{2+}, the substrate strand cleaves into two pieces, resulting in head-to-tail or tail-to-tail aggregation with a concomitant red to blue color shift with a sensing limit of 100 nM, which is unaffected by other divalent metal ions.

Aptamers are single-stranded DNA or RNA molecules that can bind target molecules with high affinity and specificity. The conformation of an aptamer usually changes upon binding to its target analyte, and this property has been used in a wide variety of sensing applications, including detection based on fluorescence intensity, polarization, energy transfer, electrochemistry or color change. Colorimetric sensors are particularly important because they minimize or eliminate the necessity of using expensive and complicated instruments. Wei et al. [432] employed unmodified gold nanoparticles as a probe for sensitive aptamer-based colorimetric sensing of proteins. In this case, the colorimetric responses for dyes adsorbed onto gold nanoparticles surface have been assessed as a means of promoting the low-potential, sensitive, and stable determination of proteins. Another sensor Liu et al; [433] developed aptamer-linked gold nanoparticles for colorimetric sensing of analytes. They have been investigated time range for the entire protocol is ~5d, including synthesis and functionalization of nanoparticles, preparation of nanoparticle aggregates and sensing.

6.1.2. Acoustic Wave Based Piezoelectric Sensors

Acoustic wave biosensors are based on the detection of mechanical acoustic waves and incorporate a biological component. Acoustic waves can be generated and received by a variety of means including piezoelectric, magnetostrictive, optical and thermal techniques [437-442]. The piezoelectric transduction effect almost exclusively has been utilized for generation and reception acoustic waves in sensor applications. Acoustic biomedical sensors are typically designed to operate in a resonant type sensor configuration implemented as an oscillator. These are mass sensitive detectors, which are operated on the basis of an oscillating crystal that resonates at a fundamental frequency. After the crystal has been coated with a biological reagent (such as an antibody) and exposed to the particular antigen a quantifiable change occurs in the resonant frequency of the crystal, which correlates to mass changes at the crystal surface. The vast majority of acoustic wave biosensors utilized piezoelectric materials as the signal transducers. Piezoelectric materials are ideal for use in this application due to their ability to generate and transmit acoustic waves in a frequency-dependent manner. The physical dimensions and properties of the piezoelectric material influence the optimal resonant frequency for the transmission of the acoustic wave.

Greatly improve sensitivity and limits of detection have measured acoustic wave biosensors. In the mass-amplified quartz crystal microbalance assay variant of this technology, antibody modified sol particles indirectly bind to an electrode surface using complexion to an analyte that has been itself captured by an antibody immobilized on the electrode surface. The large mass of the bound sol particles greatly affects the vibrational frequency of the quartz crystal and this is used as the basis for detection. The assay can be carried out in the competitive mode. The preferred diameter of sol particles is in the range of 5–100 nm. Other high-density particles (e.g. Au, Pt, CdS, Fe_3O_4, ZnO, TiO_2, polymers) may be also suitable [438-442].

6.1.3. Conductometric Biosensors

Conductometric biosensors measure the changes in the conductance of the biological component arising between a pair of metal electrode. Some researchers have constructed biosensors for estimation of glucose, urea neutral lipid/lipase and hemoglobin/pepsin by monitoring the change in the electronic conductivity arising as a change in redox potential and/or pH of the microenvironment in the polymer matrix [443-446]. Renault and his coworkers developed a new approach for immunoassays based on magnetite nanoparticles for Escherichia coli (E. coli) detection using conductometric measurements [444]. They have also investigated conductometric measurements specific test with E. coli cells and the non specific test using Staphylococcus epidermidis (S. epidermidis) and found good response as a function of antigen additions. The detection of 1 CFU/ml of E. coli induces a conductivity variation of 35 μS. Their conductometric measurements results allow detecting 500 CFU/l. Conductivity biosensors based on double-codified nanogold particles as labels were developed for hepatitis B surface antigen (HBsAg) detection. Under optimized conditions, this biosensor shows linear range as obtained 1.5–450 ng/mL HBsAg by using HRP-conjugated *anti*-HBs as secondary antibodies, while the assay sensitivity by using double-codified nanogold particles could be further increased to 0.01 ng/mL with the linear range from 0.1 to 600 ng/mL HBsAg [445].

6.1.4. Calorimetric Biosensors

The basic principle of calorimatric biosensors is that all biochemical reactions involve a change in enthalpy. Such a change in enthalpy is detected by calorimetric biosensors. Enzyme-catalyzed reactions show high selectivity and typically involve an enthalpy change, which makes them highly suitable for configuring them as thermal biosensors for a broad range of applications [446]. Danielsson in 1990 has been developed an enzyme thermister where the temperature change in an enzyme reaction was measured by a thermister device [447]. A number of substrates, enzymes and antigens have been estimated using thermister biosensors. Bhambhani et al. (2008) investigated differential scanning calorimetric studies and enzyme/protein stabilities which are bound to nanoporous inorganic materials. For detection, thermal signal generated by the kinetic reaction was measured as opposed to measuring the electrochemical signal [448].

6.1.5. Electrochemical Sensors

The conductivity and catalytic properties of metallic and semiconductor nanoparticles have been applied to electroanalytical sensing. The attachment of nanoparticles onto electrodes drastically enhances the conductivity and electron transfer from the redox analytes [450-464]. Based on this concept, Willner et al. reported several systems using nanoparticle– enzyme hybrids as electrochemical sensors [449,450]. In one example, a bioelectro-catalytic system was constructed by connecting the redox enzyme glucose oxidase (apo-GOx) onto gold nanoparticles that was functionalized with N6-(2-aminoethyl) flavin adenine (FAD). This enzyme–nanoparticle hybrid system was linked to the electrode through dithiols, or alternatively the FAD-functionalized nanoparticles were assembled onto the electrode followed by the addition of apo-GOx. This system exhibited a highly efficient electrical communication with the enhanced turnover rates as compared to native GOx and provided an effective sensor for glucose in the physiological concentration regime. An analogous electron transfer from protein to nanoparticles was used for monitoring hydrogen evolution from zinc-substituted cytochromec immobilized TiO_2 nanoparticles, as reported by Yeni Astuti et al [464]. Mirkin et al. developed a method for the detection of DNA by selective deposition of oligonucleotide-functionalized nanoparticles between two electrodes [465]. In this approach short-chain oligonucleotides were deposited onto a SiO_2 surface between two electrodes. In the presence of target DNA, oligonucleotide-functionalized gold nanoparticles hybridize on the surface. Deposition of silver by using Ag^+ salt and hydroquinone on the gold nanoparticles enhances the conductivity providing sensitivity down to 500 fM with a point mutation selectivity factor of about 100000:1 [466]. These devices are mainly based on the observation of current or potential changes due to interactions occurring at the sensor- sample matrix interface. Compared to optical methods, electrochemistry allows the analyst to work with turbid samples, and the capital cost of equipment is much lower. On the other hand, electrochemical methods present slightly more limited selectivity and sensitivity than their optical counterparts.

Recent rapid developments in biological analysis, medical diagnosis, pharmaceutical industry and environmental control fuel the urgent need for recognition of particular analyte (glucose, cholesterol, urea, lactase, DNA, antigen and pesticides detection) from samples. In this context, a significant electrochemical devices on different types of metal and metal oxide nanoparticles presently being investigated as gold nanoparticles, platinum nanoparticles, silica, carbon nanotubes (SWCNT & MWCNT), niobium oxide (Nb_2O_5), manganese oxide (MnO_2), zinc oxide (ZnO), antimony oxide (Sb_2O_3), zirconium oxide (ZrO_2), cerium oxide (CeO_2), iron oxide (Fe_3O_4), tin oxide (SnO_2), titanium oxide (TiO_2) and composite of these nanomaterials [8,14,22-24, 450-464]. Nanoparticles, possessing unique optical and electrical properties due to electron and phonon confinement effect, are receiving great deal of attention as alternative matrices for enzyme immobilization and to improve stability and sensitivity of biosensors. These interesting matrices provide high surface area for higher biomolecule loading and a compatible microenvironment helping biomolecule to retain its bioactivity. The fact that biomolecules such as enzyme, antigen or DNA exhibit dimensions comparable to those of nanoparticles (NPs) suggest that biomolecule–NPs hybrid systems may represent new materials revealing combined and synergistic properties originating from the hybrid composite. Besides this, they provide direct contact between enzyme's active site and electrode. For example, gold nanoparticles have employed for studies in biological environment because they show no surface oxidation and high biocompatibility. Xiao et al fabricated glucose biosensor based on gold nanoparticles modified dithiol SAM on Au surface [450]. They have reported that gold nanoparticles enhance the electron transfer rate by 5000 s^{-1} between enzyme's active site and electrode. Biosensors based on metal nanoparticles modified CNT electrodes lower the over potential of biochemical reaction and avoid the interference from other co-existing electroactive species. Various enzymatic, DNA and immunosensors have been reported using nanoparticles of metal, metal oxide and semiconductors.

6.1.6. Amperometric Sensors

Amperometric biosensors measure the current produced during the oxidation or reduction of a product or reactant usually at a constant applied potential. It works on the grounds of an existing linear relationship between analyte concentration and current [468-475]. The sensor potential is set at a value where the analyte, directly or indirectly, produces a current at the electrode. Nanostructured materials provide a direct redox mediator free electron communication surface between the analyte (enzymes or biomolecules) and electrode, various nanostructured materials have used for enzyme immobilization may affect significantly the response of biosensors sensitive to given species. Much work has been done in recent years on the application of nanostructured materials for the construction of amperometric biosensors. In this context, Xiao et al. (1999) have constructed a sensitive amperometric H_2O_2 biosensor wherein horse radish peroxidase (HRP) is immobilized on Au NPs supported by the thiol tailed group of the cysteamine (cyst) monolayer [473]. HRP labelled Au colloids display excellent electrocatalytic responses to the reduction of H_2O_2, increasing with the decrease in the Au colloid size. Musameh et al., showed the electrocatalytic activity of CNT toward hydrogen peroxide and NADH permitted effective low potential amperometric biosensing of glucose and ethanol in connection with the incorporation of glucose oxidase and alcohol dehydrogenase/NAD^+ within the three-dimensional CNT/Teflon matrix [474]. These advantages of CNT-based composite devices were illustrated from comparison to their graphite/Teflon counterparts, which clearly demonstrated higher sensitivity of CNT/Teflon biocomposite.

6.1.7. Potentiometric Sensors

Potentiometry is the measurement of an electrical potential difference (voltage) between two electrodes when the cell current is zero. The two electrodes are known as the working and reference electrodes. The reference electrode is required to provide a constant half-cell potential. The working electrode develops a variable potential depending on the activity or concentration of a specific analyte in solution. The change in potential is related to concentration in a logarithmic manner. The ion-selective electrode (ISE) for the measurement of electrolytes is a potentiometric technique routinely used in analytical chemistry. ISEs have also been used as transducers in potentiometric biosensors [476-482].

In recent years, several investigators have employed nanostructured materials for construction of potentiometric sensors, which have showed ultratrace level detection limits on the order of nanomolar or lower concentrations ranges. In this context, Wang and his coworkers have utilized silver ion-selective electrode as an effective transducer for sandwich immunoassays in connection to the capture and silver enlargement of gold nanoparticle tracers through potentiometer method [479]. It is anticipated that this approach may form the basis for highly sensitive bioaffinity assays. Another approach involves the use of suitably modified ion selective electrodes (ISEs) which utilise the effective transducer to detect DNA hybridization detection for capturing a secondary oligonucleotide bearing CdS-nanocrystal tags.

6.1.8. Electrochemical Impedance Spectroscopy (EIS)

Impedance spectroscopy represents a powerful method to investigate the interfacial behavior of the modified electrodes for analyte detection. In this technique, a cyclic function of small amplitude and variable frequency is applied to a transducer, and the resulting current is used to calculate the impedance at each of the frequencies probed. The amplitude of the current and potential signals and the resulting phase difference between voltage and current, which depends on the nature of the system under study, dictates the system impedance [483-498]. That the impedance has a real and an imaginary component makes its mathematical treatment quite difficult and cumbersome. The imposed signal may involve a range of frequencies and amplitudes, and the results may be interpreted according to two routes. The most rigorous approach involves solving the system of partial differential equations governing the system. The second way, which is often preferred because of its relative simplicity, consists in the interpretation of the data in terms of equivalent circuits [483-489]. The latter are made up of a combination of capacitors and resistors suitably arranged. Although this methodology is widely accepted because of ease of use, extreme care must be taken to ensure that the equivalent circuit obtained makes physical sense. An advantage of EIS compared to amperometry or potentiometry is that labels are no longer necessary, thus simplifying sensor preparation. Although impedimetric techniques are very promising, a lot of work is still needed in order to bring the technique up to a competitive level.

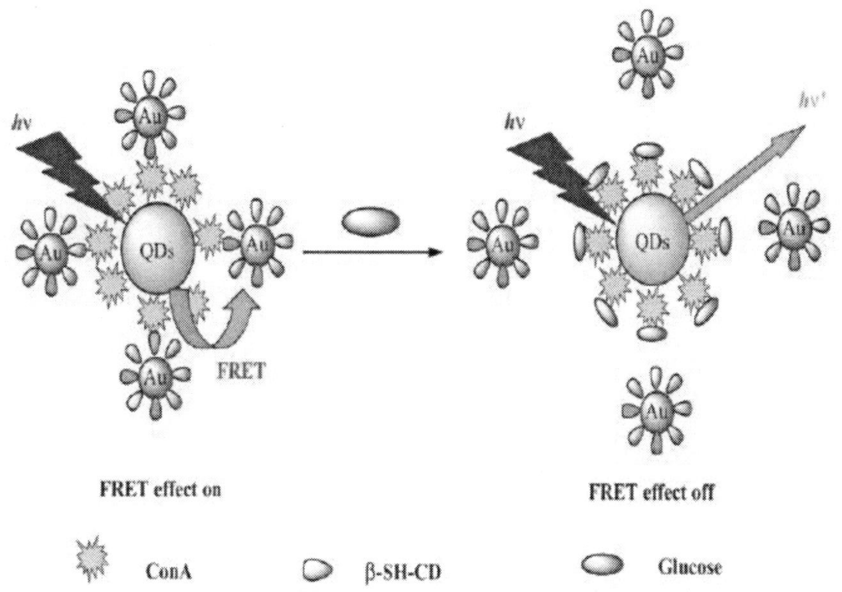

Figure 20. Chemical structure of the QDs-ConA-b-CDs-AuNPs nanobiosensor and schematic illustration of its FRET-based operating principles.

A variety of nanostructured materials have been utilized for construction of impedimetric biosensors to monitor various biological reactions at the surface of electrodes by immobilizing biomolecules such as enzymes, antibodies, nucleic acids, cell and microorganisms [490-498]. Since characteristics of the electrode, including size, shape, materials, and biological receptors greatly affects the performance of the impedimetric biosensor, modifying the electrode have received attentions. Improvements in the sensitivity of impedimetric biosensor by reducing the size of the interdigitated electrode have been reported. Nanoparticles have used as signal enhancement materials in impedimetric biosensor application due to unique electronic, photonic and catalytic properties with biomaterials [492-497]. The specific interaction between target molecules and nanoparticle-antibody conjugate induce the change of the electrical properties such as resistance and capacitance, which can be represented as the change of impedance. For instance, studies of the effect of electrode size and their separation distance have not been found in the recent literature which is still required, more investigations.

6.1.9. Surface Plasmon Resonance Sensors

SPR biosensors measure changes in refractive index caused by structural alterations in the vicinity of thin film metal surface. Current instruments operate as follows. A glass plate covered by a gold thin film is irradiated from the backside by p-polarised light (from a laser) via a hemispherical prism, and reflectivity measured as a function of the angle of incidence, θ. The resulting plot is a curve showing a narrow dip. This peak is known as the SPR minimum. The angle position of this minimum is determined by the properties of the gold-solution interface [499-509]. Hence, adsorption phenomena and even antigen–antibody reaction kinetics can be monitored using this sensitive technique (as a matter of fact, SPR is used to determine antigen–antibody affinity constants). The main drawbacks of this powerful technique lay in its complexity (specialized staff is required), high cost of equipment and large size of most currently available instruments (although portable SPR kits are also available commercially, as is the case of Texas Instruments' Spreeta system). SPR has successfully been applied to the detection of pathogen bacteria by means of immuno-reactions.

In recent years, the applications of metal nanoparticles have received significant attention because of their unique properties, such as high surface-to-volume ratios and number of available receptors. Lyon *et al.* demonstrate the characteristics of the SPR ATR-type biosensor with immobilization of colloidal gold film on the evaporated gold film, which results in a large shift and broadening of the SPR spectrum and an increase in the minimum reflectance [503,504]. Porous silica matrix functionalized with SAM proposed as a rigid matrix [504,505]. The mesoporous structure increases the surface area of the sensing layer, showing a threefold improvement over SAM on gold surface. The attachment of additional mass to the absorbed analyte receptor has also been extensively reported to amplify the change of refractive index. A wide range of nano-particles have been considered, including metal colloidal particles, latex particles, dye-doped polymer particles, or catalytic agents [499]. They are explored for various SPR approaches, from Silicon prism to fiber and waveguide structures. While the nano-particles generate pronounced SPR signals, they essentially transform an advantageous label-free sensing technique into a labeled one. Furthermore, the label, dye, or particles may be problematic in many experimental measurements studying binding affinity or biomolecular interaction, as they can influence the kinetic rate or the equilibrium-binding event. The trend is toward using nano-technology to develop sensing matrices and surfaces that are embedded the nano-particles [500].

6.1.10. Fluorescence Sensors

Fluorescence occurs when a valence electron is excited from its ground state to an excited singlet state. The excitation is produced by the absorption of light of sufficient energy. When the electron returns to its original ground state it emits a photon at lower energy [510-527]. Another important feature of fluorescence is the little thermal loss and rapid (<10 ns) light emission taking place after absorption. The emitted light is at a longer wavelength than the absorbed light since some of the energy is lost due to vibrations, this energy gap is termed Stoke's shift, and it should be large enough to avoid cross talk between excitation and emission signals. Antibodies may be conjugated to fluorescent compounds; the most common of which is fluoresce in isothiocyanate (FITC). Fluorescence resonance energy transfer (FRET) biosensors are based on the transfer of energy from a donor fluorophore to an acceptor fluorophore. Figure 20 schematically shows how this kind of biosensor works [516-518].

Significant works have been reported in the literature on utilization of nanostructured materials for construction of fluorescence sensors. In this context, semiconductors QDs used for the sensing of DNA and proteins.

Melvin and his coworkers developed a fluorescence competition assay for DNA detection using QDs and gold nanoparticles as a FRET donor–acceptor couple [517]. In the presence of complementary oligonucleotides, gold nanoparticle are released from the QD, regenerating QD fluorescence. Kim and co-workers used the similar method for sensing the avidin and glycoproteins [527]. Another paradigm for sensing relies on array-based sensing using selective recognition elements, that is, the ''chemical nose approach''. Rotello and collaborators have fabricated a sensor array by using cationic nanoparticles with various head groups and anionic poly(p-phenyleneethynylene) (PPE) fluorescent polymer [523,524]. In this sensor design, the cationic nanoparticles quench the fluorescence of the PPE polymer. Competitive binding of analyte proteins then release the PPE polymer, resulting in fluorescence restoration. Depending on the protein-nanoparticle interactions, different fluorescence response patterns were generated for individual proteins. Linear discrimination analysis (LDA) provided identification of unknown proteins, which are identified with 94.2% accuracy on the basis of 52 samples.

6.1.11. Other Sensing Methods

Surface-enhanced Raman scattering (SERS) has emerged as a powerful analytical tool that extend the possibilities of vibrational spectroscopy to solve a vast array of chemical and biochemical problems. SERS is an extension and variation of standard Raman spectroscopy, vibrational spectroscopic technique that provides detailed information about the materials at the molecular level. SERS is useful for determining molecular structural information and also provides ultrasensitive detection limits, including single molecule sensitivity; it has been used to detect pathogens that include bacteria and viruses [525-534]. There are two principle SERS configurations that have been used in biosensing, intrinsic or extrinsic, as shown in Figure 1. In intrinsic detection (Figure 21a), the analyte can be directly applied to the nanostructured surfaces and the inherent Raman spectrum of the biomolecule directly measured to identify the specimen. To allow for capture and to aid specificity of detection, antibodies, aptamers, or related molecules can be immobilized onto nanostructured surfaces as shown in Figure 21b, and the Raman spectral differences before and after capture of the specimen can be used to identify the species. In extrinsic detection, a Raman reporter molecule is used to generate a signal for detection. For example, a Au nanoparticle may be used as the SERS-active substrate to which a Raman reporter molecule is immobilized. By coating this structure with another layer of dielectrics such as SiO_2, TiO_2, or a polymer, a core-shell complex is formed in which the outer-shell may be decorated with capture molecules such as antibodies. Thus, specimens may be captured and detected via a sandwich structure as shown in Figure 21. This extrinsic SERS detection method has been successfully used for *in vivo* SERS imaging of unique or rare cancer cells [534-538].

Mirkin et al. used gold nanoparticles labeled with oligonucleotides and Raman-active dyes to achieve multiplexed detection of different DNA targets [539-543]. The SERS method was also employed for the detection of the protein–small molecule and protein–protein interactions by fabricating the nanoparticles with proteins and Raman dyes. Another ultrasensitive ''bio-barcode'' biosensor for proteins and nucleic acids has developed by Mirkin's group, where the oligonucleotide is amplified and then detected either by surface hybridization or by PCR amplification, providing a means for detection of both proteins and DNA. In case of DNA detection, a magnetic microparticle carrying partially complementary DNA to target DNA was hybridized with bio-barcoded gold nanoparticles in presence of target DNA. After the magnetic separation of the sandwich assemblies and thermal dehybridization, the barcode was released for analysis. The detection limit of this method was found at 500 zM, comparable to many PCR-based approaches. For protein detection the magnetic microparticles carry antibodies that specifically bind the target protein. The magnetic microparticle bound protein subsequently interacts with gold nanoparticles by antigen–antibody interaction. The detection limit for the chip-based sandwich hybridization process was observed at 30 aM, which was reduced using PCR amplification to 3 aM. This approach is also used for multiplexed detection of protein and DNAs by using mixture of different bio-barcoded gold nanoparticle probes.

6.2. Bioimaging, Bio-Labeling, Biomarker

A number of molecular imaging techniques such as optical/fluorescence imaging, positron emission tomography (PET), magnetic resonance imaging (MRI), X-ray computed tomography (CT), single photon emission computed tomography (SPECT), ultrasound and microwave have been employed for imaging the structure and function of systems of *in vitro* and *in vivo* biological specimens. The current development of luminescent and magnetic nanoparticles advances bioimaging technologies [544-562]. Two types of nanoparticles have been widely used for imaging: luminescent nanoprobes for optical imaging (OI) and magnetic nanoparticles for MRI. There are also dual-mode nanoparticles for simultaneous imaging by OI and MRI.

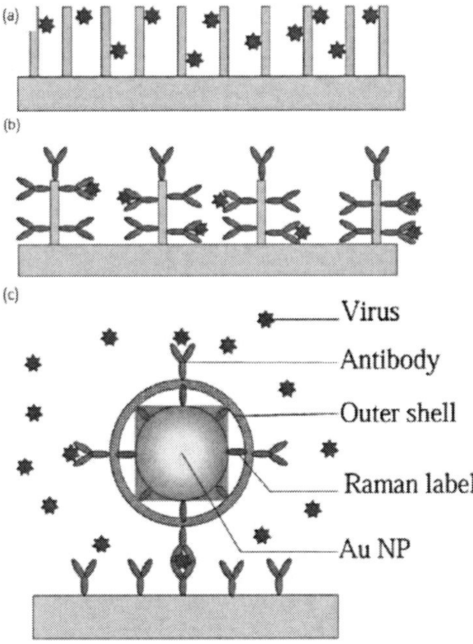

Figure 21. Different SERS detection configurations: (a) Direct intrinsic detection; (b) indirect intrinsic detection; and (c) extrinsic detection.

6.2.1. Optical Imaging

Optical imaging is an imaging technique (also known as molecular imaging) that involves inference from the deflection of light emitted from a laser or infrared source to anatomic or chemical properties of material (e.g. cell). In the area of life sciences, biotechnology and clinical diagnostics, various optical imaging, labeling or biomarker reagents, including radio-isotop, enzymes and fluorescence dyes have been used for the detection of biological molecules(such as proteins) [563-592]. Among these reagents have some drawback like as fluorescence long time stability, photo-bleaching, back ground noises caused by biological sample and analysis instrument, which has limited the effective applications of fluorescence dyes for highly sensitive bioassay. Now researchers have investigated new ways to overcome this problem by quantum dots and fluorescent nanostructured materials. These fluorescent nanostructured materials have attracted a great deal of research interest in the last decade in the field of clinical diagnostics to optically detection of bio-molecules or cells *in vivo* and *in vitro*. In particular, a very large interest in using nanostructured materials has developed due to their photo-stability and strong photo-luminescence as compared to organic dyes. As a result, they are becoming highly favored for the optically imaging or labeling of disease caused bio-molecules *in vivo* and *in vitro* diagnosis.

Most nanoparticle-based optical imaging agents can be subdivided into three categories: latex nanospheres, luminescent quantum dots (QDs) & dye-doped nanoparticle QDs and optically active luminescent nanostructure materials (such as gold, silver and luminescent lanthanide nanoparticles). The use of QDs for cell imaging was first reported by Nie's[572] and Alivisatos' groups in 1998 [573]. Compared to conventional fluorophores, QDs are photochemically stable, brighter, have a narrow, tunable and symmetric emission spectrum, and are metabolically stable. There are, however, issues of toxicity, photo-oxidation, and water solubility associated with these materials. The problem of acute toxicity and photo-oxidation can be overcome by capping with a protective shell of insulating material or semiconductor, for example, ZnS-coated CdSe core/shell QDs [579-584]. As water solubility is the key of their applications in imaging, there are long range preparation methods of QDs have been reported to make water soluble QDs and biocompatible for biological imaging, such as fabricating the surface with suitable thiolated ligand, over-coating with silica and encapsulating with amine-modified polymer. Likewise, there are large number of strategies have applied for their functionalization with organic functional groups[584,585]. Taking the advantage of the size dependent emission of QDs Choi et al. determined the size- and charge-dependent renal clearance of QDs, a very important issue for the design of biologically targeted nanoparticles for medical applications [585]. Their study revealed that zwitter ionic or neutral QDs with hydrodynamic diameter >15nm prevent renal extraction, whereas rapid and efficient elimination was observed for QDs <5.5nm in diameter. The requirement of external excitation for QDs sometimes limits their *in vivo* applications due to tissue opacity. An attractive example of a self-illuminating QD conjugate was reported by So et al. for *in vivo* imaging [586]. They overcome the issue of opacity in terms of excitation by designing a self illuminating QD that luminesces without external excitation by coupling QDs to a modified luciferase. The energy released by luciferase catalysis was transferred to QDs via resonance energy transfer resulting in emission from QDs. QD–peptide conjugates have applied to bioimaging systems [587]. A QD-peptide luminescent probe with inherent signal amplification upon interaction with a proteolytic enzyme was reported by Chang et al [588]. In this system, QDs are bound to gold nanoparticles via a proteolytically degradable peptide sequence. Upon proteolysis of the peptide linker, the QD becomes highly luminescent owing to the separation from the quenching gold NP. Other examples of bioimaging using QDs have been summarized in many recent reviews, demonstrating their utility in imaging lymph nodes and blood vessels. Rapid photo-bleaching of organic dyes is one of the biggest problems in their use as imaging agents. The optical transparent property of silica nanoparticles was used to solve this problem through encapsulation, greatly stabilizing the dyes. Silica nanoparticles can be synthesized either by a sol–gel process or by a microemulsion technique [590-592]. These dye-doped silica particles provide biocompatibility, signal amplification and low toxicity along with facile synthesis, encapsulation procedures and easy surface modification for attaching biomolecules such as proteins and peptides. Tan's group reported several procedures of covalent conjugation of biomolecules with dye-doped silica nanoparticles [593-596]. In one example, they have functionalized nanoparticles' surfaces with the enzymes glutamate dehydrogenase and lactate dehydrogenase, and used them for cell membrane staining. In another example they used aptamer conjugated magnetic and dye-doped silica nanoparticles for selective collection and detection of cancer cells.

There are, however, other fluorescent biomarkers are employed in biomedical sciences. The use of lanthanides as sources of fluorescence in luminescent assays has very recently been reviewed [597-599]. Although lanthanides pose several important advantages (good stability, low background luminescence under normal light conditions and large Stoke's shift) compared to more traditional fluorophores, their use is very restricted due to safety reasons. These nanostructured materials can be homogeneously dispersed in suitable solvents allowing exciting upconversion luminescence in transparent colloidal solutions. Upconverting nanocrystals that are dispersible in water are interesting materials for biotagging application since the excitation with NIR light strongly reduces the autofluorescence background of the biomaterials in the visible range. Compared with the traditionally used down-conversion fluorescent organic dyes and quantum dots, conceivable advantages of NIR-to-visible upconversion fluorescent bioprobes include an improved signal-to-noise ratio due to the absence of autofluorescence and reduction of light scattering and the noninvasive excitation of NIR light that falls within the "water window" (the gap in the absorption spectrum of tissue between chromophores(< 800 nm) and water (> 1200 nm)). In vivo imaging can be easily achieved because of the ability of NIR radiation to strongly penetrate tissue. Photobleaching can be greatly reduced because of the resistance to photobleaching of these inorganic particles. Multiple labeling can also be achieved by fluorescent particles with different emissions under the same excitation. In recent years, a large number of reports have been published on lanthanide (III) doped nanostructured materials because of their potential applications in proteins labeling.

6.2.2. Magnetic Resonance Imaging

Magnetic resonance imaging (MRI) is one of the most powerful non-invasive imaging modalities utilized in clinical medicine today [600-607]. MR imaging is based on the property that hydrogen protons will align and process around an applied magnetic field, B_0, which monitored by the nuclear magnetic resonance (NMR) spectrometer. Upon application of a transverse radiofrequency (rf) pulse, these protons are perturbed from B_0. The subsequent process through which these protons return to their original state is referred to as the relaxation phenomenon. Two independent processes, longitudinal relaxation (T_1-recovery) and transverse relaxation (T_2-decay), can be monitored to generate an MR image. Local variation in relaxation, corresponding to image contrast, arises from proton density as well as the chemical and physical nature of the tissues within the specimen. Exogenous contrast agents are generally introduced to enhance the tissue contrast, including complexes of Gd^{III} and magnetic nanoparticles [551,608]. Complexes of Gd^{III} in liposomes or micelles are widely used as a MRI contrast agents; however, these systems suffer from drawbacks such as Gd^{III} ion exchange with endogenous metals (e.g., Zn, Cu), and uptake of complexes in extra vascular space. Bawendi's et al have described an efficient synthetic method for magnetic and fluorescent silica microspheres fabricated by incorporating magnetic (g-Fe_2O_3) nanoparticles and CdSe/CdZnS core/shell QDs into a silica shell around preformed silica microspheres [608]. The monodisperse, cross-linked iron oxide nanoparticles reported by Weissleder et al., provide non-toxic MRI contrast agents [609-613]. These iron oxide nanoparticles are superparamagnetic in nature. The cross-linked iron oxide nanoparticles are highly stable, and convenient ''clickable'' nanoparticles have been used for targeted imaging with high cellular uptake. This superparamagnetic behavior of iron oxide nanoparticles play an important role as MRI contrasts agents, to better differentiate healthy and pathological tissues [600-606]. Recent developments in MR imaging have enabled *in vivo* imaging at near microscopic resolution. In order to visualize and track stem and progenitor cells by MR imaging, it is necessary to tag cells magnetically. Tat protein-derived peptide sequences have recently been used as an efficient way of internalizing a number of marked proteins into cells [614]. Lewin et al. hypothesized that biocompatible magnetic particles could be derivatized with similar sequences and that entire particles could be efficiently ferried into haematopoietic and neural progenitor cells in quantities upto 10–30 pg of superparamagnetic iron per cells [615]. Iron incorporation did not affect cell viability, differentiation, or proliferation of CD34+ cells. Following intravenous injection into immunodeficient mice, 4% of magnetically CD34+ cells homed to bone marrow per gram of tissue, and single cells could be detected by MRI in tissue samples. In addition, magnetically labelled cells that had homed to the bone marrow could be recovered by magneticsep aration columns. Weissleder et al. [616] presented the evidence that trans gene expression can be visualized noninvasively by MRI *in vivo*. The authors have conjugated human holo-transferrin to iron oxide nanoparticles and showed that increase in receptor levels at the cell surface can cause considerable changes in MRI signals. These superparamagnetic iron oxide nanoparticles (SPION) are relatively non-toxicwhen administered intravenously, and similar preparations are in clinical use [617], and as the iron oxide core is biodegradable, iron oxide nanoparticle degradation theoretically will allow multiple imaging of transgene expression over time.

The availability of a universal MR marker gene to image gene expression could be particularly important in monitoring gene therapy, in which exogeneous genes are introduced to ameliorate a genetic defect or to add an additional gene function to cells, and construction and testing of such vectors is currently under way. The desired strategy can also be used to image endogenous gene expression during development and pathogenesis of disease. With advances in establishing transgenic mouse models, an animal line might be developed with an imaging marker gene under the control of a given promoter under study, so that the promoter activity can be directly visualized. The work opens an exciting avenue for developing additional and complementary strategies to image gene expression in deep organs by MRI [618]. Magnetic nanoparticles have been used to detect apoptosis by MRI by Zhao et al. [619]. Apoptosis is an active process of cellular self-destruction that plays an important role in number of disorders including neurodegenerative diseases, cerebral and myocardial ischaemia and organ rejection following transplant [620]. Therapeutic treatment of tumour cells in vivo results in changes in MR image contrast that are thought to reflect the morphological features of apoptosis, such as cell shrinkage and membrane blebbing [621]. The C2 domain of synaptotagmin I, which binds to anionic phospholipids in cell membranes, was shown to bind to the plasma membrane of apoptotic cells by both flow cytometry and confocal microscopy. Administration of C2-SPION can lead to significant increases in image contrast in those regions of a tumour containing relatively large number of apoptotic cells. The authors showed that conjugation of the protein to SPION allowed detection of this binding using MRI. Specific detection of apoptotic cells using this contrast agent was demonstrated both in vitro, with isolated apoptotictumour cells and in vivo in a tumour treated with chemotherapeutic drugs [623]. The MRI technique can detect apoptosis at an early stage in the process and has the advantages over other methods such as magnetic resonance spectroscopy and radionuclide techniques, that it can detect apoptoticregions with relatively high spatial resolution. The SPION label is highly sensitive to MR detection and is also relatively non-toxic. SPION has been approved for clinical use as a blood pool agent for MRI [624].

6.3. Drug Delivery

Drug delivery (DDs) is the method or process of administering a pharmaceutical compound to achieve a therapeutic effect in humans or animals. Drug delivery technologies are patent protected formulation technologies that modify drug release profile, absorption, distribution and elimination for the benefit of improving product efficacy and safety, as well as patient convenience and compliance [625-631]. Site-specific-targeted drug delivery is important in the therapeutic modulation of effective drug dose and disease control. Targeted encapsulated drug delivery using nanoparticles is more effective for improved bioavailability, minimal side effects, decreased toxicity to other organs and is less costly. Nanoparticles based drug delivery is feasible in hydrophobic and hydrophilic states through variable routes of administration, including oral, vascular and inhalation [10, 19]. In drug delivery, several approaches are currently being tested for better site-specific delivery of an effective dose using liposomes, polymeric micelles, dendrimeres, ceramic nanoparticles, iron oxide, proteins, covalent binding, adsorption, conjugation, and encapsulation methods. Extended circulation of liposomes with entrapped doxorubicin was reported to be 300-fold more effective with better pharmacokinetic ability than free doxorubicin in the treatment of Kaposi's sarcoma and metastatic cancer. Kumar et al. reported that nanoparticle surfaces modified with cationic chitosan were efficient for drug delivery both *in vitro* and *in vivo*[631]. The cationic surface of nanoparticles facilitated the penetration through cell membrane, and the payload release was triggered by intracellular glutathione, relying on the ca. 1000-fold higher intracellular concentration of GSH relative to the extracellular environment. Gelperina et al. reported that in chemotherapy treatment for tuberculosis, nanoparticles based drug delivery improved drug bioavailability, reduced dose frequency and overcame the nonadherence problem encountered in the control of tuberculosis epidemics [632]. The major disadvantage of most chemotherapies is that they are relatively non-specific. The therapeutic drugs are administered intravenously leading to general systemic distribution, resulting in deleterious side-effects as the drug attacks normal, healthy cells in addition to the target tumour cells. For example, the side effects of anti-inflammatory drugs on patients who have chronic arthritis can lead to the discontinuation of their use. However, if such treatments could be localized, e.g. to the site of a joint, then the continued use of these very potent and effective agents could be made possible.

Very recently, Bhatia and his coworkers designed multifunctional superparamagnetic nanoparticles for remote release of bound drugs [633]. The particles transduce external electromagnetic force (EMF) at 350–400 kHz to local heating for breaking hydrogen bonds between DNA chains. Besides the surface chemistry of nanoparticles, the unique physical properties of nanoparticles have been utilized in the design of DDSs. Covalently functionalized magnetic nanoparticles (Fe_3O_4) with doxorubicin (DOX), an anticancer drug, through an acid-labile hydrazone linker [634]. The carrier was then encapsulated with thermo-sensitive polymer for temperature-controlled release of the drug. The hybrid system released DOX efficiently in mild acidic buffer solution of pH 5.3. Polizzi et al. have likewise shown that (nitric oxide) can be efficiently released at acidic pH from gold nanoparticles [635]. Neuman et al. designed water-soluble nanocontainer for nitric oxide storage based on electrostatic assembly of dihydrolipoic acid-coated quantum dots and cationic dinitro complexes that uses energy transfer from the core to release nitric oxide [636]. In another approach, doping of Ag/Au nanoparticles serves as an antenna to absorb the energy from a laser beam of ''biologically friendly'' near-infrared (NIR) region, causing local heating and disruption of microcapsules.

An exciting potential use of nanotechnology in cancer treatments is the exploration of tumor-specific thermal scalpels to heat and burn tumors. O'Neal et al. observed in mice that selective photothermal ablation of tumor using near infrared-absorbing polyethylene-coated gold nanoshells of 130 nm inhibited tumor growth and enhanced survival of animals for up to 90 days compared with controls [637]. Perkel et al. reported that antibody-coated magnetic iron nanoparticles were effective to heat and literally cook the tumors [638]. In similar work performed in athymic mice using antibody-coated iron NPs, DeNardo et al. showed specific targeted binding to tumors and tumor necrosis within 24 hr after therapy with better response [639]. The efficacy of different antibodies conjugated to nanoparticles, including transferrin and epidermal growth factor receptor, was examined in animal studies [639,640]. Kaul and Amiji have observed that tumor-targeted gene delivery using polyethylene glycol–modified gelatin nanoparticles was highly effective, biocompatible, biodegradable, and long circulating for systemic delivery to solid tumors [641]. From these studies, it is apparent that nanotechnology will profoundly affect human health through advances in medicine, science, and industry. The potential human benefits of nanotechnology are innumerable and include many aspects of human life with wide a variety of products.

Chapter 7

7. CONCLUSION AND FUTURE PROSPECTS

In this book we reviewed some basic concepts regarding the nanotechnology and nanoscience including various new types of nanostructures such as nanowires, nanobelts, nanorods, nanotube and nanodisks and shown how these nanostructured materials to be synthesized and characterized according to the current appropriate biomedical application. The unique physical, chemical and optical properties of nanostructure materials such as high surface area, strong adsorption capability, biocompatibility, non-toxicity, chemical and mechanical stability, isoelectric point, electrical conductivity, catalytic efficiency (oxygen storage capacity), and reduction in potential are likely to be helpful in healthcare systems. Nanostructured materials classified according to the morphology of the nanoparticles and novel properties of the nanostructured materials have employed in broad range applications in biomedical sciences including biosensors (enzymatic biosensors, DNA biosensor, pH sensor and immunosensors), drug delivery and bioimaging techniques. We have focused in particular on nanostructured materials: classification, properties, fabrication methods, characterization techniques and many biomedical advantages of nanomaterials that are currently being explored. The surface and core properties of these systems can be engineered for individual and multimodal applications, including biomolecular recognition, therapeutic delivery, biosensing and bioimaging. Nanoparticles have already been used for a wide range of applications both *in vitro* and *in vivo*. Full realization of their potential, however, requires addressing a number of open issues, including acute and long-term health effects of nanomaterials as well as scalable, reproducible manufacturing methods and reliable metrics for characterization of these materials. There are opportunities in this respect for more interdisciplinary approaches, for example, to ensure that the laboratory based experiments can more explicitly emulate the expected conditions that would be encountered *in vivo*. There is also scope for significant contributions via the mathematical modelling of complex systems, with the objective of understanding more specifically the full gamut of physical phenomena and effects that together determine whether, in the final analysis, a given application will be successful.

References

[1] Poole, CP Jr.; Owen,s, F. J. *Introduction to Nanotechnology Wiley*, 2003.

[2] Shipway, AN; Katz, E; Willner, I. *Chempyschem*, 2008, 1, 18-52.

[3] Murray, RW. *Chem. Rev.*, 2008, 108, 2688-2720.

[4] Reiss, P; Protiere, M; Li, L. *Small*, 2009, 5, 154-168.

[5] Caruthers, SD; Wickline, SA; Lanza, GM. *Current Opinion in Biotechnology*, 2007, 18, 26-30.

[6] Engel, E; Michiardi, A; Navarro, M; Lacroix, D; Planell, J. A. *Trends in Biotechnology*, 2007, 26, 39-47.

[7] Zhang, L; Webster, T. J. *Nano Today*, 2009, 4, 66-80.

[8] Fortina, P; Kricka, LJ; Surrey, S; Grodzinski, P. *Trends in Biotechnology*, 2005, 23, 168-173.

[9] Liu, H; Webster, T. J. *Biomaterials*, 2007, 28, 354-369.

[10] Byrappa, K; Ohara, S; Adschiri, T. *Adv. Drug Deliv. Rev.*, 2008, 60, 299-327.

[11] De, M; Ghosh, PS; Rotello, VM. *Adv. Mater*, 2008, 20, 1-17.

[12] Jain, KK. *Clinica Chimica Acta*, 2005, 358, 37-54.

[13] Bawarski, WE; Chidlowsky, E; Bharali, DJ; Mousa, SA. *Nanomedicine*, 2008, 4, 273-282.

[14] Carrascosa, LG; Moreno, M; Alvarez, M; Lechuga LM. *Trends in Analytical Chemistry*, 2006, 25, 196-206.

[15] Kerman, K; Saito, M; Yamamura, S; Takamura, Y; Tamiya, E. *Trands Analytical Chem.*,2008, 27, 585-592.

[16] Ferrari, M. *Nat. Rev. Cancer*, 2005, 5, 161-71.

[17] Gwinn, MR; Vallyathan, V. *Environmental Health Perspectives*, 2006, 14, 1818-1825.

[18] Sahoo, SK; Labhasetwar, V. *Drug Delivery Today*, 2003, 8, 1112-1120.
[19] Jianrong, C; yuqing M; Nongyue, H; Xiaohua, W; Sijiao, L. *Biotech. Advances*, 2004, 22, 505-518.
[20] Sokolova, V; Epple, M. *Angew Chem. Int. Ed*, 2008, 47, 1382-1395.
[21] Jamieson, T; Bakhshi, R; Petrova, D; Pocock, R; Imani, M; Seifalian, AM. *Biomaterials*, 2007, 28, 4717-4732.
[22] Pandey, P; Datta, M; Malhotra, BD. *Analytical Letters*, 2008, 41, 159-209.
[23] Allen, BL; Kichambare, PD; Star, A. *Adv. Mater*, 2007, 19, 1439-1451.
[24] Mornet, S; Vasseur, S; Grasset, F; Veverka, P; Goglio, G; Demourgues, A; Portier, J; Pollert, E; Duguet, E. *Progress in Solid State Chemistry*, 2006, 34, 237-247.
[25] Archakov, AI; Ivanov, YD. *Mol. BioSyst*, 2007, 3, 336-342.
[26] Kuchibhatla, SVNT; Karakoti, AS; Bera, D; Seal S. *Prog. Mater. Science*, 2007, 52, 699-913.
[27] Valentini, F; Palleschi, G. *Analytical Letters*, 2008, 41, 479-520.
[28] Caruso, F. *Adv. Mater*, 2001, 13, 11-22.
[29] Xia, Y; Xiong, Y; Lim, B; Skrabalak, SE. *Angew. Chem. Int. Ed*, 2009, 48, 60-103.
[30] Taniguchi, N. *On the Basic Concept of nanotechnology*, *ICPE*, 1974.
[31] Hussain, F; Hojjati, M; Okamoto, M; Gorga, RE; *J. Composite Mater*, 2006, 40, 1511-1575.
[32] Hatchett, D. W; Josowicz, M. *Chem. Rev.*, 2008, 108,746-769.
[33] Malinauskas, A; Malinauskiene, J; Ramanavicius, A. *Nanotechnology*, 2005, 16, R51-R62.
[34] Rajesha, Ahuja, T; Kumar, D. *Sensors and Actuators B*, 2009, 136, 275-286.
[35] Zou, H; Wu, S; Shen, J. *Chem. Rev.*, 2008, 108, 3893-3957.
[36] Baibarac, M; Gómez-Romero, P. *J. Nanosci. Nanotech*, 2006, 6, 1-14.
[37] Cushing, BL; Kolesnichenko, VL; O'Connor, C. *J. Chem. Rev.*, 2004, 104, 3893-3946.
[38] Zhang, X; Manohar, SK. *Chem. Commun.*, 2004, 2360-2361.
[39] Nastase, C; Nastase, F; Vaseashta, A; Stamatin, I. *Prog. Solid State Chem.*, 2006, 34, 181-189.
[40] Tseng, RJ; Huang, J; Ouyang, J; Kaner, RB; Yang, Y. *Nano Lett*, 2005, 5, 1077-1080.
[41] Luo, X; Killard, AJ; Smyth, MR. *Chem. Eur. J*, 2007, 13, 2138-2143.
[42] Yan, Y; Yu, Z; Huang, Y; Yuan, W; Wei, Z. *Adv. Mater*, 2007, 19, 3353-3357.

[43] Dai, H. *Acc. Chem. Res.*, 2002, 35, 1035-1044.
[44] Kam, NWS; O'Connell, M; Wisdom, JA; Dai, H. *PNAS*, 2005, 102, 11600-11605.
[45] Georgakilas, V; Kordatos, K; Prato, M; Guldi, DM; Holzinger, M; Hirsch A. *J. Am. Chem. Soc.*, 2002, 124, 760-761.
[46] Kim, SN; Rusling, JF; Papadimitrakopoulos F. *Adv. Mater*, 2007, 19, 3214-3228.
[47] Wang, W; Fernando, KAS; Lin, Y; Meziani, MJ; Veca, LM; Cao, L; Zhang, P; Kimani, MM; Sun, YP. *J. Am. Chem. Soc.*, 2008, 130, 1415-1419.
[48] Yang, R; Jin, J; Chen, Y; Shao, N; Kang, H; Xiao, Z; Tang, Z; Wu, Y; Zhu, Z; Tan, W. *J. Am. Chem. Soc.*, 2008, 130, 8351-8358.
[49] Guo, S; Wang, E. *Anal. Chim. Acta*, 2007, 598, 181-192.
[50] Wiley, BJ; Chen, Y; McLellan, JM; Xiong, Y; Li, ZY; Ginger, D; Xia,Y. *Nano. Lett*, 2007, 7, 1032-1036.
[51] Thiel, J; Pakstis, L; Buzby, S; Raffi, M; Ni, C; Pochan, DJ; Shah, SI. *Small*, 2007, 3, 799-803.
[52] Aslan, K; Wu, M; Lakowicz, JR; Geddes, CD. *J. Am. Chem. Soc.*, 2007, 129, 1524-1525.
[53] Ghilane, J; Fan, FRF; Bard, AJ; Dunwoody, N. *Nano Lett*, 2007, 7, 1406-1412.
[54] Du, J; Han, B; Liu, Z; Liu, Y; Kang, DJ. *Crystal Growth & Design*, 2007, 7, 900-904.
[55] Lesniak, W; Bielinska, AU; Sun, K; Janczak, KW; Shi, X; Baker Jr., JR; Balogh, LP. *Nano Lett*, 2005, 5, 2123-2130.
[56] Han, MS; Jean, AKRL; Mirkin, CA. *J. Am. Chem. Soc.*, 2006, 128, 4954-4955.
[57] Thaxton, CS; Georganopoulou, DG; Mirkin, CA. *Clin. Chim. Acta*, 2006, 363, 120-126.
[58] Xue, C; Mirkin, CA. *Angew. Chem. Int. Ed*, 2007, 46, 2036-2038.
[59] Pan, Y; Neuss, S; Leifert, A; Fischler, M; Wen, F; Simon, U; Schmid, G; Brandau, W; Dechent, W. J. *Small*, 2007, 3, 1941-1949.
[60] Pyrpassopoulos, S; Niarchos, D; Nounesis, G; Boukos, N; Zafiropoulou, I; Tzitzios, V. *Nanotechnology*, 2007, 18, 485604.
[61] Nie, LB; Yang, Y; Li, S; He, NY. *Nanotechnology*, 2007, 18, 305501.
[62] Debouttière, PJ; Roux, S; Vocanson, F; Billotey, C; Beuf, O; Réguillon, AF; Lin, Y; Rostaing, SP; Lamartine, R; Perriat, P; Tillement, O. *Adv. Funct. Mater*, 2006, 16, 2330-2339.
[63] Jana, NR; Ying, JY. *Adv. Mater*, 2008, 20, 430-434.

[64] Elder, A; Yang, H; Gwiazda, R; Teng, X; Thurston, S; He, H; Oberdörster, G. *Adv. Mater*, 2007, 19, 3124-3129.
[65] Chen, J; Xiong, Y; Yin, Y; Xia, Y. *Small*, 2006, 2, 1340-1343.
[66] Wang, C; Daimon, H; Onodera, T; Koda, T; Sun, S. *Angew. Chem. Int. Ed*, 2008, 47, 3588-3591.
[67] Ren, J; Tilley, RD. *Small*, 2007, 3, 1508-1512.
[68] Xiong, Y; Xia, Y. *Adv. Mater*, 2007, 19, 3385-3391.
[69] Das, K; Nagarajan, V; Goswami, MLN; Panda, D; Dhar, A; Ray, SK. *Nanotechnology*, 2007, 18, 095704.
[70] Xie, ZQ; Chen, D; Li, ZH; Zhao, YY; Lu, M. *Nanotechnology*, 2007, 18, 115716.
[71] Jensen, JS; Buttenschön, DA; Pedersen, TPL; Chevallier, J; Nielsen, BB; Larsen, AN. *J. Appl. Phys.*, 2007, 101, 056108.
[72] Bian, LF; Zhang, CG; Chen, WD; Hsu, CC; Shi, T. *Appl. Phys. Lett*, 2006, 89, 231927.
[73] Elfström, N; Karlström, AE; Linnros, J. *Nano Lett*, 2008, 8, 945-949.
[74] Li, C; Huang, H; Yang, S; Zheng, R; Yang, W; Liu, Z; Ringer, S. *Mater.Lett*, 2009, 63, 1016-1018.
[75] Shehata, F; Fathy, A; Abdelhameed, M; Moustafa, SF. *Materials & Design*, 2009, 30, 2756-2762.
[76] Ansari, AA; Kaushik, A; Solanki, PR; Malhotra, BD. *Electrochem. Commun*, 2008, 10, 1246-1249.
[77] Ansari, AA; Solanki, PR; Malhotra, BD. *Appl. Phys. Lett*, 2008, 92, 263901-3.
[78] Chen, X; Li, G; Su, Y; Qiu, X; Li, L; Zou, Z. *Nanotechnology*, 2009, 20, 115606.
[79] Kaushik, A; Solanki, PR; Ansari, AA; Ahmad, S; Malhotra, BD. *Nanotechnology*, 2009, 20, 055105.
[80] Ghoshal, T; Kar, S; De, SK. *Appl. Surf. Sci.*, 2009, doi:10.1016/j.apsusc.2009.05.021.
[81] Askarinejad, A; Morsali, A. *Chem. Engin. J*, 2009, 150, 569-571.
[82] Zhuo, L; Ge, J; Cao, L; Tang, B. *Crystal Growth & Design*, 2009, 9, 1-6.
[83] Chun-Hong Kuo and Michael, H. Huang, *J. Phys. Chem. C*, 2008, 112, 18355-18360.
[84] Bao, Q; Li, CM; Liao, L; Yang, H; Wang, W; Ke, C; Song, Q; Bao, H; Yu, T; Loh, KP; Guo, J. *Nanotechnology*, 2009, 20, 065203.
[85] Chaudhari, NK; Yu, JS. *J. Phys. Chem. C*, 2008, 112, 19957-19962.
[86] Na, HB; Song, IC; Hyeon, T. *Adv. Mater*, 2009, 21, 1-16.
[87] Gupta, AK; Gupta, M. *Biomaterials*, 2005, 26, 3995-4021.

[88] Sun, C; Lee, JSH; Zhang, M. *Adv. Drug Delivery Rev.*, 2008, 60, 1252-1265.

[89] Chertok, B; Moffat, BA; Davida, AE; Yua, F; Bergemann, C; Ross, BD; Yang, VC. *Biomaterials*, 2008, 29, 487-496.

[90] Kumar, A; Singhal, A. *Nanotechnology*, 2007, 18, 475703.

[91] An, K; Kwon, SG; Park, M; Na, HB; Baik, SI; Yu, JH; Kim, D; Son, JS; Kim, YW; Song, IC; Moon, WK; Park, HM; Hyeon, T. *Nano Lett.*, 2008, 8, 4252-4258.

[92] Zeng, H; Sun, S. *Adv. Funct. Mater*, 2008, 18, 391-400.

[93] Li, Z; Tan, B; Allix, M; Cooper, AI; Rosseinsky, M. J. *Small*, 2008, 4, 231-239.

[94] Geng, BY; Ma, JZ; You, JH. *Crystal Growth & Design*, 2008, 8, 1443-1447.

[95] Park, JH; Maltzahn, G; Zhang, L; Schwartz, MP; Ruoslahti, E; Bhatia, SN; Sailor, MJ. *Adv. Mater*, 2008, 20, 1630-1635

[96] Kim, J; Rong, C; Lee, Y; Liu, JP; Sun, S. *Chem. Mater*, 2008, 20, 7242-7245.

[97] Thunemann, AF; Rolf, S; Knappe, P; Weidner, S. *Anal. Chem.*, 2009, 81, 296-301

[98] Liu, JC; Tsai, PJ; Lee, YC; Chen, YC. *Anal. Chem.*, 2008, 80, 5425-5432.

[99] Larsen, BA; Haag, MA; Serkova, NJ; Shroyer, KR; Stoldt, CR. *Nanotechnology*, 2008, 19, 265102.

[100] Sljukic, B; Banks, CE; Compton, RG. *Nano Lett*, 2006, 6, 1556-1558.

[101] Liu, Z; Wang, J; Xie, D; Chen, G. *Small*, 2008, 4, 462-466.

[102] Rossi, LM; Quach, AD; Rosenzweig, Z. *Anal. Bioanal. Chemistry*, 2004, 380, 606-613.

[103] Qiu, J; Peng, H; Liang, R. *Electrochemistry Commun*, 2007, 9, 2734-2738.

[104] Qu, S; Wang, J; Kong, J; Yang, P; Chen, G. *Talanta*, 2007, 71, 1096-1102.

[105] Zhao, G; Xu, JJ; Chen, HY. *Electrochem. Commun*, 2006, 8, 148-154.

[106] Cao, D; Hu, N. *Biophys. Chem.*, 2006, 121, 209-217.

[107] Zhang, HL; Zou, XZ; Lai, GS; Han, DY; Wanga, F. *Electroanalysis*, 2007, 19, 1869-1874.

[108] Lin, MS; Leu, H. J. *Electroanalysis*, 2005, 17, 2068-2073.

[109] Hrbac, J; Halouzka, V; Zboril, R; Papadopoulos, K; Triantis, T. *Electroanalysis*, 2007, 19, 1850-1854.

[110] Kaushik, A; Solanki, PR; Ansari, AA; Ahmad, S; Malhotra, BD. *Electrochem. Commun*, 2008, 10, 1364-1368.
[111] Liao, MH; Guo, JC; Chen, WC. *J. Magn. Magnetic Mater*, 2006, 304, e421-e423.
[112] Sinha, G; Patra, A. *Chem. Phys. Lett*, 2009, 473, 151-154.
[113] Lucovsky, G; Hinkle, CL; Fulton, CC; Stoute, NA; Seo, H; Lüning, J. *Rad. Phys. Chem.*, 2006, 75, 2097-2101.
[114] Zhou, Y; Zhang, H; Xue, M; Wu, C; Wu, X; Fu, Z. *J. Power Sour*, 2006, 162, 1373-1378.
[115] George, PP; Gedanken, A. *Eur. J. Inorg. Chem.*, 2008, 919-924.
[116] Lu, W; Liu, Q; Sun, OZ; He, J; Ezeolu, C; Fang, J. *J. Am. Chem. Soc.*, 2008, 130, 6983-6991.
[117] Yang, J; Li, C; Quan, Z; Kong, D; Zhang, X; Yang, P; Lin, J. *Crys.Growth & Desig*, 2008, 8, 695-699.
[118] Vomiero, A; Bianchi, S; Comini, E; Faglia, G; Ferroni, M; Sberveglieri, G. *Cryst. Growth & Design*, 2007, 7, 2500-2504.
[119] Singh, N; Zhang T; Lee, PS. *Nanotechnology*, 2009, 20, 195605-7.
[120] Capobianco, JA; Vetrone, F; Boyer, JC; Speghini, A; Bettinelli, M. *Optic.Mater*, 2002, 19, 259-268.
[121] Chesnokov, VV; Bedilo, AF; Heroux, DS; Mishakov, IV; Klabunde, KJ. *J. Catalysis*, 2003, 218, 438-446.
[122] Choudary, BM; Ranganath, KVS; Yadav, J; Kantam, ML. *Tetrah. Lett*, 2005, 46, 1369-1371.
[123] Hsu, CS; Chan, CC; Huang, HT; Peng, CH; Hsu, WC. *Thin Solid Films*, 2008, 516, 4839-4844.
[124] Zhan, JH; Zhang, ZD; Qian, XF; Wang, C; Xie, Y; Qian, YT. *Mater. Res. Bull*, 1999, 34, 497-501.
[125] Han, YF; Chen, F; Zhong, ZY; Ramesh, K; Widjaja, E; Chen, LW. *Cataly. Commun*, 2006, 7, 739-744.
[126] Fei, J; Cui, Y; Yan, X; Qi, W; Yang, Y; Wang, K; He, Q; Li, J. *Adv. Mater*, 2008, 20, 452-456.
[127] Weixin, Z; Cheng, W; Xiaoming, Z; Yi, X; Yitai, Q. *Solid State Ion*, 1999, 117, 331-335.
[128] Ai, Z; Zhang, L; Kong, F; Liu, H; Xing, W; Qiu, J. *Mater.Chem.Phys.*, 2008, 111, 162-167.
[129] Zhao, N; Nie, W; Liu, X; Tian, S; Zhang, Y; Ji, X. *Small*, 2008, 4, 77-81.
[130] Qian, Y; Xie, YT; Wang, WZ; Zhang, SY; Zhang, YH. *Science*, 1996, 272, 1926-1927.

[131] Wang, X; Li,Y. *J. Am. Chem.Soc.*, 2002, 124, 2880-2881.

[132] Liu, JF; Li, XL; Li, YD. *J.Crys.Growth*, 2003, 247, 419-424.

[133] Shanker, V; Samal, SL; Pradhan, GK; Narayana, C; Ganguli, AK. *Solid State Sci.*, 2009, 11, 562-569.

[134] Wang, H; Jiao, X; Chen, D. *J. Phys. Chem. C*, 2008, 112, 18793-18797.

[135] Deraz, NM; Selim, MM; Ramadan, M. *Mater.Chem.Phys.,* 2009, 113, 269-275.

[136] Hirano, M; Ito, T. *J. Solid State Chem.*, 2009, 182, 1581-1586.

[137] Wang, J; Zhong, S; Wang, GX; Bradhurst, DH; Ionescu, M; Liu, HK; Dou, SX. *J. Alloys Comp.*, 2001, 327, 141-145.

[138] Shrestha, S; Mills, CE; Lewington, J; Tsang, SC. *J. Phys. Chem.*, *B* 2006, 110, 25633-25637.

[139] Shrestha, S; Yeung, CMY; Mills, CE; Lewington, J; Tsang, SC. *Angew. Chem. Intern. Ed*, 2007, 46, 3855-3860.

[140] Yan, L; Yu, R; Liu, G; Xing, X. *Scripta Mater*, 2008, 58, 707-710.

[141] Nayak, J; Sahu, SN. *Mater. Lett*, 2007, 61, 1388-1391.

[142] Hu, Y; Zhang, H; Yang, H. *J.Alloys Comp.*, 2007, 428, 327-331.

[143] Zhang, X; Neiner, D; Wang, S; Louie, AY; Kauzlarich, SM. *Nanotechnology*, 2007, 18, 095601.

[144] Arriagada, FJ; Asare, KO. *J. Coll. Inter. Sci.*, 1999, 211, 210-220.

[145] Warren, SC; Disalvo, FJ; Wiesner, U. *Nature Materials*, 2007, 6, 156-161.

[146] Hessel, CM; Henderson, EJ; Veinot, JGC. *Chem. Mater*, 2006, 18, 6139-6146.

[147] Liu, Z; Yang, Q; Zhang, H; Wang, L; Li, D; Yang, D. *Nanotechnology*, 2008, 19, 165601.

[148] Dattoli, EN; Wan, Q; Guo, W; Chen, Y; Pan, X; Lu, W. *Nano Lett*, 2007, 7, 2463-2469.

[149] Ng, SH; Santos, DI; Chew, SY; Wexler, D; Wang, J; Dou, SX; Liu, HK. *Electrochem. Commun*, 2007, 9, 915-919.

[150] Park, MS; Kang, YM; Wang, GX; Dou, SX; Liu, HK. *Adv. Funct. Mater*, 2008, 18, 455-461.

[151] Li, LL; Zhang, WM; Yuan, Q; Li, ZX; Fang, CJ; Sun, LD; Wan, LJ; Yan, CH. *Cryst. Growth & Design*, 2008, 8, 4165-4172.

[152] Yan, L; Yu, R; Chen, J; Xing, X. *Crystal Growth & Design*, 2008, 8, 1474-1477.

[153] Zhu, J; Lu, Z; Aruna, ST; Aurbach, D; Gedanken, A. *Chem. Mater*, 2000, 12, 2557-2566.

[154] Jiang, L; Sun, G; Zhou, Z; Sun, S; Wang, Q; Yan, S; Li, H; Tian, J; Guo, J; Zhou, B; Xin, Q. *J. Phys. Chem., B*, 2005, 109, 8774-8778.

[155] Ibarguen, CA; Mosquera, A; Parra, R; Castro, MS; Paez, JER. *Mater.Chem.Phys.*, 2007, 101 433-440.

[156] Robles, LIV; Campero, A. *J. Phys. Chem. C*, 2008, 112, 19930-19933.

[157] Kumar, A; Singh, P; Kulkarni, N; Kaur, D. *Thin Solid Films*, 2008, 516, 912-918.

[158] Ng, SH; Chew, SY; Wang, J; Wexler, D; Tournayre, Y; Konstantinov, K; Liu, HK. *J. Power Sources*, 2007, 174, 1032-1035.

[159] Bahgat, M; Paek, MK; Pak, JJ. *J. Alloys Comp.*, 2009, 472, 314-318.

[160] Djaoued, Y; Priya, S; Balaji, S. *J. Non-Crystalline Solids*, 2008, 354, 673-679.

[161] Deepa, M; Singh, DP; Shivaprasad, SM; Agnihotry, SA. *Current Appl. Phys.*, 2007, 7, 220-229.

[162] Li, J; Seok, SI; Chu, B; Dogan, F; Zhang, Q; Wang, Q. *Adv. Mater*, 2009, 21, 217-221.

[163] Huang, J; Huang, Z; Guo, W; Wang, Ml; Cao, Y; Hong, MC. *Crystal Growth & Design*, 2008, 8, 2444-2446.

[164] Li, Q; Zhang, J; Liu, B; Li, M; Yu, S; Wang, L; Li, Z; Liu, D; Hou, Y; Zou,Y; Zou, B; Cui, T; Zou, G. *Crystal Growth & Design*, 2008, 8, 1812-1814.

[165] Liu, R; Ren, Y; Shi, Y; Zhang, F; Zhang, L; Tu, B; Zhao, D. *Chem. Mater*, 2008, *20*, 1140-1146.

[166] Kumar, A; Jose, R; Fujihara, K; Wang, J; Ramakrishna, S. *Chem. Mater*, 2007, 19, 6536-6542.

[167] Sun, WT; Yu, Y; Pan, HY; Gao, XF; Chen, Q; Peng, LM. *J. Am. Chem. Soc.*, 2008, 130, 1124-1125.

[168] Kniprath, R; Duhm, S; Glowatzki, H; Koch, N; Rogaschewski, S; Rabe, JP; Kirstein, S. *Langmuir*, 2007, *23*, 9860-9865.

[169] Kuo, CY; Lu, SY. *Nanotechnology*, 2008, 19, 095705.

[170] Luo, M; Cheng, K; Weng, W; Song, C; Du, P; Shen, G; Xu, G; Han, G. *Nanotechnology*, 2009, 20, 215605.

[171] Woan, K; Pyrgiotakis G; Sigmund, W. *Adv. Mater*, 2009, 21, 1-7.

[172] Pighini, C; Aymes, D; Millot N; Saviot, L. *J. Nanoparticle Res.*, 2007, 9, 309-315.

[173] Wang, KP; Teng, H. *Appl. Phys. Lett*, 2007, 91, 173102,

[174] Mack, JM; Tsuchiya, H; Schmuki, P. *Angew. Chem. Int. Ed*, 2005, 44, 2100-2102.

[175] Zhang, XJ; Ma, TY; Yuan, ZY. *Eur. J. Inorg. Chem.*, 2008, 2721-2726.

[176] Pagano, JJ; Jr, TB; Steinbock, O. *Angew. Chem. Int. Ed*, 2008, 47, 9900-9903.

[177] Baolin Wang, Jijun Zhao, Jianming Jia, Daning Shi, Jianguo Wan, Wang, G. *Appl. Phys. Lett*, 2008, 93, 021918.

[178] Hartlieb, KJ; Raston, CL; Saunders, M. *Chem. Mater.*, 2007, 19, 5453-5459.

[179] Wang, M; Kim, EJ; Hahn, SH; Park, C; Koo, KK. *Crystal Growth & Design*, 2008, 8, 501-506.

[180] Wu, LYL; Tok, AIY; Boey, FYC; Zeng, XT; Zhang, XH. *IEEE Transaction Nanotechnology*, 2007, 6, 497-503.

[181] Shen, L; Bao, N; Yanagisawa, K; Gupta, A; Domen, K; Grimes, CA. *Crystal Growth & Design*, 2007, 7, 2742-2748.

[182] Umar, A; Karunagaran, B; Kim, SH; Suh, EK; Hahn, YB. *Inorg. Chem.*, 2008, 47, 4088-4094.

[183] Garcia, MA; Merino, JM; Pinel, EF; Quesada, A; dela Venta, J; Gonzalez, MLR; Castro, GR; Crespo, P; Llopis, J; Calbet, JM. G; Hernando, A. *Nano Lett*, 2007, 7, 1489-1494.

[184] Lu, F; Cai, W; Zhang, Y. *Adv. Funct. Mater*, 2008, 18, 1047-1056.

[185] Irimpan, L; Nampoori, VPN; Radhakrishnan, P; Krishnan, B; Deepthy, A. *J. Appl. Phys.*, 2008, 103, 033105.

[186] Huang, Z; Zheng, X; Yan, D; Yin, G; Liao, X; Kang, Y; Yao, Y; Huang, D; Hao, B. *Langmuir*, 2008, 24, 4140-4144.

[187] Dong, WS; Lin, FQ; Liu, CL; Li, MY. *J.Colloid Inter. Sci.*, 2009, 333, 734-740.

[188] Zhou, S; Garnweitner, G; Niederberger, M; Antonietti, M. *Langmuir*, 2007, 23, 9178-9187.

[189] Clavel, G; Willinger, MG; Zitoun, D; Pinna, N. *Eur. J. Inorg. Chem.*, 2008, 863-868.

[190] Li, W; Huang, H; Li, H; Zhang, W; Liu, H. *Langmuir*, 2008, 24, 8358-8366.

[191] Kumari, L; Li, W; Wang, D. *Nanotechnology*, 2008, 19, 195602.

[192] O'Donnell, A; Yach, K; Reven, L. *Langmuir*, 2008, 24, 2465-2471.

[193] Zong, S; Cao, Y; Zhou, Y. Ju, H. *Biosens. Bioelectron*, 2007, 22, 1776-1782.

[194] Zhang, L; Han, Y; Lu, J. *Nanotechnology*, 2008, 19, 165706.

[195] Sun, XL; Tok, AIY; Lim, SL; Boey, FYC; Kang, CW; Ng, HW. *J. Appl. Phys.*, 2008, 103, 034308.

[196] Chen, GY; Liu, Y; Zhang, YG; Somesfalean, G; Zhang, ZG; Sun Q; Wang, FP. *Appl. Phys. Lett*, 2007, 91, 133103.

[197] Feng, J; Shan, G; Maquieira, A; Koivunen, ME; Guo, B; Hammock, BD; Kennedy, IM. *Anal. Chem.*, 2003, 75, 5282-5286.

[198] Huang, CC; Su, CH; Li, WM; Liu, TY; Chen, JH; Yeh, CS. *Adv. Funct. Mater*, 2009, 19, 249-258.

[199] Fortin, MA; Petoral Jr, RM; Soderlind, F; Klasson, A; Engstrom, M; Veres, T; Kall, PO; Uvdal, K. *Nanotechnology*, 2007, 18, 395501.

[200] Norek, M; Kampert, E; Zeitler, U; Peters, JA. *J. Am. Chem. Soc.*, 2008, 130, 5335-5340.

[201] Goldys, EM; Tomsia, KD; Jinjun, S; Dosev, D; Kennedy, IM; Yatsunenko, S; Godlewski, M. *J. Am. Chem. Soc.*, 2006, 128, 14498-14505.

[202] Losurdo, M; Giangregorio, MM; Capezzuto, P; Bruno, G; Toro, RG; Malandrino, G; Fragalà, IL; Barreca, L; Tondello, E; Suvorova, AA; Yang, D; Irene, EA. *Adv. Funct. Mater*, 2007, 17, 3607-3612.

[203] Liu, C; Liu, J; Dou, K. *J. Phys. Chem.*, *B* 2006, 110, 20277-20281.

[204] Cui, F; Zhang, J; Cui, T; Liang, S; Ming, L; Gao, Z; Yang, B. *Nanotechnology*, 2008, 19, 065607.

[205] Mahajan, SV; Dickerson, JH. *Nanotechnology*, 2007, 18, 325605.

[206] Liu, J; Li, Y. *Adv. Mater*, 2007, 19, 1118-1122.

[207] Zhu, L; Li, J; Li, Q; Liu, X; Meng, J; Cao, X. *Nanotechnology*, 2007, 18, 055604.

[208] Deligne, N; Gonze, V; Bayot, D; Devillers, M. *Eur. J. Inorg. Chem.*, 2008, 896-902.

[209] Wang, F; Xue, X; Liu, X. *Angew. Chem. Int. Ed*, 2008, 47, 906-909.

[210] Riwotzki, K; Haase, M. *J. Phys. Chem.*, *B*, 1998, 102, 10129-10135.

[211] Fan, W; Bu, Y; Song, X; Sun, S; Zhao, X. *Crystal Growth Des*, 2007, 7, 2361-2366.

[212] Qian, L; Zhu, J; Chen, Z; Gui, Y; Gong, Q; Yuan, Y; Zai, J; Qian, X. *Chem. Eur. J*, 2009, 15, 1233-1240.

[213] Gu, M; Liu, Q; Mao, S; Mao, D; Chang, C. *Crystal Growth & Design*, 2008, 8, 1422-1425.

[214] Huignard, A; Buissette, V; Franville, AC; Gacoin, T; Boilot, JP. *J. Phys. Chem. B*, 2003, 107, 6754-6759.

[215] Yan, R; Li, Y. *Adv. Func. Mater*, 2005, 15, 763-770.

[216] Sivakumar, S; Veggel, FCJM; van Raudsepp, M. *J. Am. Chem. Soc.*, 2005, 127, 12464-12465.

[217] Wang, F; Zhang, Y; Fan, X; Wang, M. *Nanotechnology*, 2006, 17, 1527-1532.

[218] Sudarsan, V; Veggel, FCJM. van Herring, RA; Raudsepp, M. *J. Mater.Chem.*, 2005, 15, 1332-1342.

[219] Wang, X; Zhuang, J; Peng, Q; Li, Y. *Inorg. Chem.*, 2006, 45, 6661-6665.

[220] Zhu, L; Liu, X; Meng, J; Cao, X. *Crystal Growth & Design*, 2007, 7, 2505-2511.

[221] Darbandi, M; Nann, T. *Chem. Commun*, 2006, 776-778.

[222] Zeng, JH; Su, J; Li, ZH; Yan, RX; Li, YD. *Adv. Mater*, 2005, 17, 2119-2123.

[223] Wang, F; Zhang, Y; Fanc, X; Wang, M. *J. Mater. Chem.*, 2006, 16, 1031-1034.

[224] Wang, J; Hu, J; Tang, D; Liua X; Zhen, Z. *J. Mater. Chem.*, 2007, 17, 1597-1601.

[225] Wang, ZL; Quan, ZW; Jia, PY; Lin, CK; Luo, Y; Chen, Y; Fang, J; Zhou, W; O'Connor, CJ; Lin, J.*Chem. Mater,* 2006, 18, 2030-2037.

[226] Sivakumar, S; Diamente, PR; Veggel, FCJM. van *Chem. Eur. J*, 2006, 12, 5878-5884.

[227] Li, C; Quan, Z; Yang, J; Yang, P; Lin, J. *Inorg. Chem.*, 2007, 46, 6329-6337.

[228] Nuñez, NO; Míguez, H; Quintanilla, M; Cantelar, E; Cussó, F; Ocaña, M. *Eur. J. Inorg. Chem.*, 2008, 4517-4524.

[229] Yi, G; Lu, H; Zhao, S; Ge, Y; Yang, W; Chen, D; Guo, LH. *Nano Lett*, 2004, 4, 2191-2196.

[230] Wang, L; Li, P; Li, Y. *Adv. Mater*, 2007, 19, 3304-3307.

[231] Boyer, JC; Cuccia, LA; Capobianco, JA. *Nano Lett*, 2007, 7, 847-852.

[232] Zhuang, J; Liang, L; Sung, HHY; Yang, X; Wu, M; Williams, ID; Feng, S; Su, Q. *Inorg.Chem.*, 2007, 46, 5404-5410.

[233] Yi, GS; Chow, GM. *Adv. Funct. Mater*, 2006, 16, 2324-2329.

[234] Li, C; Yang, J; Quan, Z; Yang, P; Kong, D; Lin, J. *Chem. Mater*, 2007, 19, 4933-4942.

[235] Xun, X; Feng, S; Wang, J; Xu, R. *Chem. Mater*, 1997, 9, 2966-2968.

[236] Wei, Y; Lu, F; Zhang, X; Chen, D. *Chem. Mater*, 2006, 18, 5733-5737.

[237] Zhang, F; Wan, Y; Yu, T; Zhang, F; Shi, Y; Xie, S; Li, Y; Xu, L; Tu; Zhao, D. *Angew. Chem. Int. Ed*, 2007, 46, 7976 -7979.

[238] Shan, J; Qin, X; Yao, N; Ju, Y. *Nanotechnology*, 2007, 18, 445607.

[239] Zeng, JH; Li, ZH; Su, J; Wang, L; Yan, R; Li, Y. *Nanotechnology*, 2006, 17, 3549-3555.

[240] Wang, L; Li, Y. *Chem. Mater*, 2007, 19, 727-734.

[241] Sun, Y; Chen, Y; Tian,L; Yu, Y; Kong, X; Zhao, J; Zhang, H. *Nanotechnology*, 2007, 18, 275609.
[242] Ptacek, P; Schäfer, H; Kömpe, K; Haase, M. *Adv. Funct. Mater*, 2007, 17, 3843-3848.
[243] Liang, L; Xu, H; Su, Q; Konishi, H; Jiang, Y; Wu, M; Wang, Y; Xia, D. *Inorg. Chem*., 2004, 43, 1594-1596.
[244] Yi, GS; Chow, GM; *Chem. Mater*, 2007, 19, 341-343.
[245] Schafer, H; Ptacek, P; Kompe, K; Haase, M. *Chem. Mater*, 2007, 19, 1396-1400.
[246] Li, Z; Zhang, Y; Jiang, S. *Adv. Mater*, 2008, 20, 4765-4769.
[247] Schafer, H; Ptacek, P; Zerzouf, O; Haase, M. *Adv. Funct. Mater*, 2008, 18, 2913-2918.
[248] Liang, X; Wang, X; Zhuang, J; Peng, Q; Li, Y. *Adv. Funct. Mater*, 2007, 17, 2757-2765.
[249] Chen, G; Sun, S; Zhao, W; Xu, S; You, T. *J. Phys. Chem. C*, 2008, 112, 20217-20221.
[250] Lehmann, O; Meyssamy, H; Ko1mpe, K; Schnablegger, H; Haase, M. *J. Phys. Chem., B*, 2003, 107, 7449-7453.
[251] Zharkouskaya, A; Feldmann, C; Trampert, K; Heering, W; Lemmer, U. *Eur. J. Inorg. Chem*., 2008, 873-877.
[252] Di, W; Willinger, MG; Ferreira, RAS; Ren, X; Lu, S; Pinna, N. *J. Phys. Chem. C*, 2008, 112, 18815-18820.
[253] Li, H; Zhu, G; Ren, H; Li, Y; Hewitt, IJ; Qiu, S. *Eur. J. Inorg. Chem*., 2008, 2033-2037.
[254] Zhang, YW; Yan, ZG; You, LP; Si, R; Yan, CH. *Eur. J. Inorg. Chem*., 2003, 4099-4104.
[255] Heer, S; Lehmann, O; Haase, M; Gudel, HU. *Angew. Chem. Int. Ed*, 2003, 42, 3179-3182.
[256] Lehmann, O; Kompe, K; Haase, M. *J. Am. Chem. Soc.,* 2004, 126, 14935-14942.
[257] Stouwdam, JW; Hebbink, GA; Huskens, J; Veggel, FCJM. van *Chem. Mater*, 2003, 15, 4604-4616.
[258] Hebbink, GA; Stouwdam, JW; Reinhoudt, DN; van Veggel, FCJM. *Adv. Mater*, 2002, 14, 1147-1150.
[259] Kang, SH; Bozhilov, KN; Myung, NV; Mulchandani, A; Chen, W. *Angew. Chem. Int. Ed,* 2008, 47, 5186-5189.
[260] Jun, YW; Choi, JS; Cheon, J. *Angew. Chem. Int. Ed*. 2006, 45, 3414-3439.

[261] Rogach, AL; Gaponik, N; Lupton, JM; Bertoni, C; Gallardo, DE; Dunn, S; Pira, NL; Paderi, M; Repetto, P; Romanov, SG; Dwyer, CO; Torres, CMS; Eychmller, A. *Angew. Chem. Int. Ed*, 2008, 47, 6538- 6549.

[262] Scholes, GD. *Adv. Funct. Mater*, 2008, 18, 1157-1172.

[263] Wang, H; Qi, L. *Adv. Funct. Mater*, 2008, 18, 1249-1256.

[264] Jamieson, T; Bakhshi, R; Petrova, D; Pocock, R; Imani, M; Seifalian, AM. *Biomaterials*, 2007, 28, 4717-4732.

[265] Zheng, Y; Gao, S; Ying, JY. *Adv. Mater*, 2007, 19, 376-380.

[266] Gao, X; Cui, Y; Levenson, RM; Chung, LWK; Nie, S. *Nature Biotechnology*, 2004, 22, 969-976.

[267] Bidoz, SF; Jennings, TL; Klostranec, JM; Fung, W; Rhee, A; Li, D; Chan, WCW. *Angew. Chem. Int. Ed.*, 2008, 47, 5577-5581.

[268] Chan, WCW; Nie, S. *Science*, 1998, 281, 2016-2018.

[269] Jiang, W; Papa, E; Fischer, H; Mardyani, S; Chan, WCW. *Trends Biotechnology*, 2004, 22, 607-609.

[270] Fischer, HC; Liu, L; Pang, KS; Chan, WCW. *Adv. Funct. Mater*, 2006, 16, 1299-1305.

[271] Klostranec, JM; Chan, WCW. *Adv. Mater*, 2006, 18, 1953-1964.

[272] Mamedova, NN; Kotov, NA; Rogach, AL; Studer, J. *Nano Lett*, 2001, 1, 281-286.

[273] Kirchner, C; Liedl, T; Kudera, S; Pellegrino, T; Javier, AM; Gaub, HE; Sto1lzle, S; Fertig, N; Parak, W. J. *Nano Lett,* 2005, 5, 331-338.

[274] Steckel, JS; Zimmer, JP; Sullivan, SC; Stott, NE; Bulovic, V; Bawendi, MG. *Angew. Chem. Int. Ed*, 2004, 43, 2154-2158.

[275] Lim, J; Jun, S; Jang, E; Baik, H; Kim, H; Cho, J. *Adv. Mater*, 2007, 19, 1927-1932.

[276] Unnia, C; Philip, D; Smith, SL; Nissamudeen, KM; Gopchandran, KG. *Spectrochim. Acta Part A*, 2009, 72, 827-832.

[277] Seo, H; Kim, SW. *Chem. Mater*, 2007, 19, 2715-2717.

[278] Conroy, J; Byrne, SJ; Gunko, YK; Rakovich, YP; Donegan, JF; Davies, A; Kelleher, D; Volkov, Y. *Small*, 2008, 4, 2006-2015.

[279] Asokan, S; Krueger, KM; Colvin, VL; Wong, MS. *Small*, 2007, 3, 1164-1169.

[280] Sun, SQ; Li, T.*Crystal Growth & Design*, 2007, 7, 2367-2371.

[281] Salata, OV. *J.Nanobiotechnology*, 2004, 2, 1-6.

[282] Sinha, R; Kim, GJ; Nie, S; Shin, DM. *Mol. Cancer Ther.*, 2006, 5, 1909-1917.

[283] Ganta, S; Devalapally, H; Shahiwala, A; Amiji, M. *J. Contr. Release*, 2008, 126, 187-204.

[284] Kim, KY. *Nanomed.Nanotech. Biol.Med*, 2007, 3, 103-110.
[285] Gaoa, J; Xu, B. *Nano Today*, 2009, 4, 37-51.
[286] Azzazy, HME; Mansour, MMH. *Clin. Chim. Acta*, 2009, 403, 1-8.
[287] Liu, H; Webster, T. J. *Biomaterials*, 2007, 28, 354-369.
[288] Pankhurst, QA; Connolly, J; Jones, SK; Dobson, J. *J. Phys. D: Appl. Phys.*, 2003, 36, R167-R181.
[289] Riehemann, K; Schneider, SW; Luger, TA; Godin, B; Ferrari, M; Fuchs, H. *Angew. Chem. Int. Ed.*, 2009, 48, 872-897.
[290] Brigger, I; Dubernet, C; Couvreur, P. *Adv. Drug Delivery Rev.*, 2002, 54, 631-651.
[291] Yang, J; Gunn, J; Dave, SR; Zhang, M; Wang, YA; Gao, X. *Analyst*, 2008, 133, 154-160.
[292] Meng, L; Fu, C; Lu, Q. *Prog. Natural Sci.*, 2009, 19, 801-810.
[293] Hoppe, HA. *Angew. Chem. Int. Ed.*, 2009, 48, 3572-3582.
[294] Klostranec, JM; WCW. Chan, *Adv. Mater*, 2006, 18, 1953-1964.
[295] Guimard, NK; Gomez, N; Schmidt, CE. *Prog. Polym. Sci.*, 2007, 32, 876-921.
[296] Capek, I. *Adv.Colloid Inter. Science*, 2004, 110, 49-74.
[297] Gleiter, H. *Progress in Materials Science*, 1989, 33, 223-315.
[298] Boylan, K; Ostrander, D; Erb, U; Palumbo, G; Aust, KT. *Scrip. Metallurg. Mater*, 1991, 25, 2711-2716.
[299] McCandlish, LE; Kear, BH; Kim, BK. *Nanostructured Mater*, 1992, 1, 119-124.
[300] Gleiter, H. *Nanostructured Mater*, 1992, 1, 1-19.
[301] Siegel, RW. *Physics of New Materials* (Ed. Fujita, F.E.), Springer-Verlag Berlin Heidelberg, Chap. 1994, 4.
[302] Goodenough, JB; Manthiram, A; Paranthaman, M; Zhen, YS. *Solid State Ionics*, (M., Balanski, Eds. et al.), *Elsevier*, Amsterdam, 1992, 79-90.
[303] Etsell, TH; Flengas, *Chem. Rev.*, 1970, 70, 339-376.
[304] Ashcroft, NW; Mermin, ND. *Solid State Phys.* Saunders Collage Publishing, 1976, 2-11, 616-623.
[305] Chiang, YM. *Mass and Charge Transport in Ceramics*, 79-89.
[306] Chiang, YM; Lavik, EB; Kosacki, I; Tuller, HL; Ying, JY. *J. Electroceramics*, 1997, 1, 7-14.
[307] Chiang, YM; Lavik, EB; Blom, DA. *Nanostr. Mat*, 1997, 9, 633-642.
[308] Gleiter, H. *Z. Metallkd.* 1995, 86, 78-83.
[309] Giannelis, EP; Mehtora, V; Vassiliou, JK; Shull, RD; MacMichael, RD; Ziolo, RF. *Nanophase Materials* (GC; Hadjipanayis, & RW. Siegel, Eds.) Kluwer Academic Publishers, the Netherlands, 1994, 197-204.

[310] Brus, L. *J. Phys. Chem.*, 1994, 98, 3575-3581.

[311] Kittel, C. *Introduction to Solid State Physics*, John Wiley and Sons, Inc. 1992, 183-187.

[312] Wilson, WL; Szajowski, PF; Brus, LE. *Science*, 1993, 262, 1242-1244.

[313] Brus, LE. *Nanophase Materials* (GC; Hadjipanayis, RW. Siegel, Eds.) Kluwer Academic Publishers, the Netherlands, 1994, 433-448.

[314] Ashcroft, NW; Mermin, ND. *Solid State Physics*, Saunders Collage Publishing, 1976, 718-722.

[315] Wohlfarth, EP. *Ferromagnetic Mater.* North-Holland Publishing Company, Amsterdam, 1980, 2, 436-444.

[316] Mallinson, JC. *The Foundations of Magn. Recording*, Academic Press, San Diego, USA, 1987, 28-30.

[317] Cullity, *Introduction to Magnetic Mater.*Addison Wesley Publishing Company, Reading, USA, 1972, 410-418.

[318] Gangopadhyay, S; Hadjipanayis, GC; Sorensen, CM; Klabunde, KJ. *Mater. Res.Soc. Sym. Proc.*, 1993, 286, 55-60.

[319] Gangopadhyay, S; Hadjipanayis, GC; Dale, B; Sorensen, CM; Klabunde, KJ; Papaefthymiou, V; Kostikas, A. *Phys. Rev. B*, 1992, 45, 9778-9787.

[320] Uchikoshi, T; Sakka, Y; Yoshitake, M; Yoshihara, K. *Nanostructured Mater*, 1994, 4, 199-206.

[321] Yiping, L; Hadjipanayis, GC; Papaefthymiou, V; Soerensen, CM; Klabunde, KJ. *Nanophase Materials* (GC; Hadjipanayis, RW. Siegel, Eds.) Kluwer Academic Publishers, the Netherlands 1994, 625-633.

[322] Somorjai, GA. *Pure & Appl. Chem.*, 1978, 50, 963-969.

[323] Klabunde, KJ; Li, YX; Khaleel, A. *Nanophase Mater.* (Eds. Hadjipanayis, G.C.and Siegel, R.W.) Kluwer Academic Publishers, the Netherlands, 1994, 757-769.

[324] Somorjai, GA. *Surface Chem. and Cataly.*John Wiley & Sons, Inc. New York, 1994.

[325] Liu, Z; Tabora, J; Davis, RJ. *J. Catal*, 1994, 149, 117-126.

[326] Tschöpe, A; Ying, JY. *Nanophase Materials*, (GC; Hadjipanayis, & RW. Siegel, Eds.) Kluwer Academic Publishers, the Netherlands, 1994, 781-784.

[327] Ying, JY. *Nanophase Materials* (GC; Hadjipanayis, RW. Siegel, Eds.) Kluwer Academic Publishers, the Netherlands, 1994, 197-204.

[328] Conner, WC; Jr; Falconer, JL. *Chem. Rev.*, 1995, 95, 759-788.

[329] Vaarkamp, M; Miller, JT; Modica, FS; Koningsberger, DC. *J. Catal*, 1996, 163, 294-305.

[330] Singh, SP; Arya, SK; Pandey, P; Malhotra, BD; Saha, S; Sreenivas, K; Gupta, V. *Appl. Phys. Lett*, 2007, 91, 063901-3.
[331] Nakamura, T; Minoura, H; Muto, H. *Thin Solid Films*, 2002, 405, 109-116.
[332] Sun, Y; Fuge, GM; Ashfold, MNR. *Chem. Phys. Lett*, 2004, 396, 21-26.
[333] Sun, Y; Fuge, GM; Ashfold, MNR. *Superlatt. & Microstruct.* 2006, 39, 33-40.
[334] Bae, CH; Park, SM; Park, SC; Ha, HS. *Nanotechnology*, 2006, 17, 381-384.
[335] Zheng, B; Lu, C; Gu, G; Makarovski, A; Finkelstein, G; Liu, J. *Nano Lett*, 2002, 2, 895-898.
[336] Lieber, CM; Duan, XF. *Adv. Mater*, 2000, 12, 298-302.
[337] Wu, JJ; Liu, SC. *Adv. Mater*, 2002, 14, 215-218.
[338] Haga, K; Kamidaira, M; Kashiwaba, Y; Sekiguchi, T; Watanabe, H. *J.Crystal Growth*, 2000, 214-215, 77-80.
[339] Jain, S; Kodas, TT; Smith, MH. *Chem.Vapor Deposition*, 1998, 4, 51-59.
[340] Muthukumar, S; Sheng, H; Zhong, J; Zhang, Z; Emanetoglu, NW; Lu, Y. *IEEE Transactions on Nanotechnology*, 2003, 2, 50-54.
[341] Kim, SW; Fujita, S; Fugita, S. *Appl. Phys. Lett*, 2005, 86, 015231.
[342] Zhang, BP; Binh, NT; Wakatuki, K; Segawa, Y; Yamada, Y; Usami, N; Kawasaki, M; Koinuma, H. *Appl. Phys. Lett*, 2004, 84, 4098-4100.
[343] Nakahara, K; Takasu, H; Fons, P; Iwata, K; Yamada, A; Matsubara, K; Hunger, R; Niki, S. *J. Crystal Growth*, 2001, 227-228, 923-928.
[344] Fonoberov, VA; Baladin, AA. *Appl. Phys. Lett*, 2004, 85, 5971-5973.
[345] George, SM; Ott, AW; Klaus, JW. *J. Phys. Chem.*, 1996, 100, 13121-13131.
[346] Scharrer, M; Wu, X; Yamilov, A; Cao, H; Chang, RPH. *Appl. Phys. Lett*, 2005, 86, 1-3.
[347] Hoivik, N; Elam, JW; George, SM; Gupta, KC; Bright, VM; Lee, YC. *Microwave Symposium Digest,* 2002 *IEEE MIT-S International*, 2002, 1129-1232.
[348] Ansari, AA; Sumana, G; Pandey, MK; Malhotra, BD. *J. Mater. Res*, 2009, 24, DOI: 10.1557/JMR.2009.0212.
[349] Ansari, AA; Kaushik, A; Solanki, PR; Malhotra, BD. *Electroanalysis*, 2009, 21, 965-972.
[350] Solanki, PR; Kaushik, A; Ansari, AA; Malhotra, BD. *Appl.Phys. Lett*, 2009, 94, 143901-3.
[351] Ansari, AA; Solanki, PR; Malhotra, BD. *Sensor Lett*, 2009, 7, 64-71.

[352] Solanki, PR; Kaushik, A; Ansari, AA; Tiwari, A; Malhotra, BD. *Sens. & Actuat. B*, 2009, 137, 727-735.

[353] Ansari, AA; Singh, R; Sumana, G; Malhotra, BD. *Analyst*, 2009, 13, 997-1002.

[354] Zhu, JJ; Wang, H. *In Encyclopedia of Nanoscience and Nanotechnology*, HS. Nalwa, (Ed.); American Scientific Publishers: New York 2003.

[355] Ghosh, M; Lawes, G; Gayen, A; Subbanna, GN; Reiff, WM; Subramanian, MA; Ramirez, AP; Zhang, JP; Seshadri, R. *Chem. Mater*, 2004, 16, 118-124.

[356] Vinodgopal, K; He, Y; Ashokkumar, M; Grieser, F. *J. Phys. Chem., B*, 2006, 110, 3849-3852.

[357] Fujimoto, T; Teravachi, S; Umehara, H; Kojima, I; Henderson, W. *Chem. Mater*, 2001, 13, 1057-1060.

[358] Okitsu, K; Xue, A; Tanabe, S; Matsumoto, H; Yobiko, Y. *Langmuir*, 2001, 17, 7717-7720.

[359] Barbour, K; Ashokkumar, M; Caruso, RA; Grieser, F. *J. Phys. Chem. B*, 1999, 103, 9231-9236.

[360] Caruso, RA; Ashokkumar, M; Grieser, F. *Langmuir*, 2002, 18, 7831-7836.

[361] Okitsu, K; Ashokkumar, M; Grieser, F. *J. Phys. Chem. B*, 2005, 109, 20673-20675.

[362] Dokoutchaeei, A; James, JT; Koene, SC; Pathak, S; Prakash, GKS; Thompson, ME. *Chem. Mater*, 1999, 11, 2389-2399.

[363] Breen, ML; Dinsmore, AD; Pink, RH; Qadri, SB; Ratna, BR. *Langmuir*, 2001, 17, 903-907.

[364] Pol, VG; Srivastava, DN; Palchik, O; Palchik, V; Slifkin, MA; Weiss, AM; Gedanken, A. *Langmuir*, 2002, 18, 3352-3357.

[365] Park, JE; Atobe, M; Fuchigami, T. *Chem. Lett*, 2005, 34, 96-97.

[366] Wang, X; Zhuang, J; Peng Q; Li, YD. *Nature*, 2005, 437, 121-124.

[367] Wen, B; Liu C; Liu, Y. *Inorg. Chem.*, 2005, 44, 6503-6505.

[368] Ahmad, A; Senapati, S; Khan, MI; Kumar, R; Sastry, M. *Langmuir*, 2003, 19, 3550-3553.

[369] Kim, CS; Moon, BK; Park, JH; Choi, BC; Seo, HJ. *J. Cryst. Growth*, 2003, 250, 309.

[370] Wen, B; Liu C; Liu, YJ. *J. Phys. Chem. B*, 2005, 109, 12372.

[371] Lee, HQ; Chua, SJ; Loh, KP; Fitzgerald, EA; Koh, YW. *Nanotechnology*, 2006, 17, 483-488.

[372] Fang, Y; Wen, X; Yang, S; Pang, Q; Ding, L; Wang, J; Ge, W. *J. Sol-Gel Sci.Tech.*, 2005, 36, 227-234.

[373] Wang, JX; Sun, XW; Wei, A; Lei, Y; Cai, XP; Li, CM; Dong, ZL. *Appl. Phys. Lett*, 2006, 88, 233106-3.

[374] Wei, A; Sun, XW; Wang, JX; Lei, Y; Cai, XP; Li, CM; Dong, ZL; Huang, W. *Appl. Phys. Lett*, 2006, 89, 123902-3.

[375] Pérez, MA; Moiraghi, R; Coronado, EA; Macagno, VA. *Crystal Growth & Design*, 2008, 8, 1377-1383.

[376] Rouanet, A; Pichelin, G; Roucau, C; Snoeck, E; Monty. C. Nanophase Materials. Kluwer Academic Publishers, *Dordrecht*, The Netherlands, 1994.

[377] Stokenius, W; Schulamn JH; Prince, LM. J. *Phys. Chem.*, 1959, 63, 1677.

[378] Quintela, MAL; Rivas. J. *J. Colloid Interface Sci.*, 1993, 158, 446-451.

[379] Lodhi, AK; Robinson, BH; Towey, T; Hermann, C; Knoche, W. Thesing. U. The structure, *dynamics and equilibrium properties of colloidal systems*, Kulwer Acad. Publ. Dordrecht, 1990, 324.

[380] Seip, CT; O'Connor. R. J. *Nanostruct. Mater*, 1999, 12,183.

[381] Ingelsten, HH; Bagwe, R; Palmqvist, A; Skoglundh, M; Svanberg, C; Holmberg. K. *J. Colloid Interface Sci.*, 2001, 241,104-111.

[382] Barnickel, P; Wokaun, A; Sager, W; Eicke. HF. *J. Colloid Inter. Sci.*, 1992, 148, 80-90.

[383] Manna, A; Kulkarni, BD. *Chem. Mater*, 1997, 9, 3032-3036.

[384] Dutta P; Fendler. J.H. *J. Colloid Interface Sci.* 2002, 247, 47-53.

[385] Iida, M; Ohkawa, S; Er, H; Asaoka, N; Yoshikawa. H. *Chem. Lett*, 2002, 1050.

[386] Hamada, K; Hatanaka, K; Kawai, T; Konno. K. *Shikizai Kyokaishi*, 2000, 73, 385.

[387] Karpov, SV; Popov, AK; Slatko, VV; Shevnina. GB. *Colloid J*, 1988, 57, 199.

[388] Lufimpadio, N; Nagy, JB; Derouane. EG. *Surfactant in solution.* Plenum, New York, 1984, 3.

[389] Chen, DH; Wu, SH. *Chem. Mater*, 2002, 12, 1354-1360.

[390] Nagy, JB. *Colloid Surf*, 1989, 35, 201-220.

[391] Cheng, B; Samulski, ET. *J. Mater. Chem.*, 2001, 11, 2901-2902.

[392] Hernandez, BA; Chang, KS; Fisher, ER; Dorhout, PK. *Chem. Mater*, 2002, 14, 480-482.

[393] Chen, J; Xu, L; Li, W; Gou, X. *Adv. Mater*, 2005, 17, 582-586.

[394] Lakshmi, BB; Dorhout, PK; Martin, CR. *Chem. Mater*, 1997, 9, 857-862.

[395] Shi, K; Chi, Y; Yu, H; Xin, B; Fu, H. *J. Phys. Chem. B*, 2005, 109, 2546-2551.

[396] Yang, H; Shi, Q; Tian, B; Lu, Q; Gao, F; Xie, S; Fan, J; Yu, C; Tu, B; Zhao, D. *J. Am.Chem. Soc.*, 2003, 125, 4724-4725.

[397] Caruso, RA; Antonietti, M. *Chem. Mater*, 2001, 13, 3272-3282.

[398] Ajayan, PM; Stephan, O; Redlich, P; Colliex, C. *Nature*, 1995, 375, 564-567.

[399] Satishkumar, BC; Govindaraj, A; Nath, M; Rao, CNR. *J. Mater. Chem.*, 2000, 10, 2115-2119.

[400] Rao, CNR; Satishkumar, BC; Govindaraj, A. *Chem. Comm.*, 1997, 1581-1582.

[401] Min, YS; Bae, EJ; Jeong, KS; Cho, YJ; Lee, JH; Choi, WB; Park, GS. *Adv.Mater*, 2003, 15, 1019-1022.

[402] Kumar, CSSR; Hormes, J; Leuschner, C. *Nanofabrication Towards Biomedical Appl.* Eds., Wiley-VCH: Weinheim, 2005.

[403] Reetz, MT; Helbig, W. *J. Am. Chem. Soc.*, 1994, 116, 7401-7402.

[404] Becker, JA; Schäfer, R; Festag, R; Ruland, W; Wendorff, JH; Pebler, J; Quaiser, SA; Helbig, W; Reetz, MT. *J. Chem. Phys.*, 1995, 103, 2520.

[405] Reetz, MT; Helbig, W; Quaiser, SA. *Chem. Mater*, 1995, 7, 2227.

[406] Reetz, MT; Quaiser, SA. *Angew. Chem. Int. Ed.*, 1995, 34, 2240.

[407] Xu, J; Jia, C; Cao, B; Zhang, WF. *Electrochim. Acta*, 2007, 52, 8044-8047.

[408] Chandera, R; Raychaudhuri, AK. *Solid State Commun.*, 2008, 145, 81-85.

[409] Li, GR; Dawa CR; Bu, Q; Zhen, Fl. Lu, XH; Ke, ZH; Hong, HE; Yao, CZ; Liu, P; Tong, YX. *Electrochem. Commun,* 2007, 9, 863-868.

[410] Chang, ST; Leu, IC; Hon, MH. *Electrochem. Solid-State Lett*, 2002, 5, C71-C74.

[411] Chang, M; Cao, XL; Zeng, H; Zhang, L. *Chem. Phys. Lett*, 2007, 446, 370-373.

[412] Limmer, SJ; Cao, G. *Adv. Mater*, 2003, 15, 427-431.

[413] Zhang, X; Wang, H; Scattergood, RO; Narayan, J; Koch, CC. *Acta Mater*, 2002, 50, 3527-3533.

[414] El-Sherik, M; Erb, U. *J. Mater. Sci.*, 1995, 30, 5743-5749.

[415] Koch, CC; Cho, YS. *Nanostruct. Mater*, 1992, 1, 207-213.

[416] Eckert, J; Hozler, JC; Krill, CE; Johnson, WL. *J. Mater. Res. Soc.*, 1992, 7, 1751-1761.

[417] Fecht, HJ; Hellstern, E; Fu, Z; Johnson, WL. *Metallurg. Transac*, 1990, 21A, 2333-2337.

[418] Park, MS; Kang, YM; Wang, GX; Dou, SX; Liu, HK. *Adv. Funct. Mater*, 2008, 18, 455-461.
[419] Robert, CL; Long, JW; Pettigrew, KA; Stroud, RM; Rolison, DR. *Adv. Mater*, 2007, 19, 1734-1739.
[420] Flores, LLD; Bon, RR; Galvan, AM; Prokhorov, E; Hernandez, JG. *J. Phys. Chem.Solids*, 2003, 64, 1037-1042.
[421] Sawicka, KM; Prasad, AK; Gouma, PI. *Sensor Lett*, 2005, 3, 1-5.
[422] Huang, XJ; Choi, YK. *Sens. and Actuators B*, 2007, 122, 659-671.
[423] Ghadiali, JE; Stevens, MM. *Adv. Mater*, 2008, 9999, 1-5.
[424] Klostranec, JM; Chan, WCW. *Adv. Mater*, 2006, 18, 1953-1964.
[425] Willner, I; Baron, R; Willner, B. *Adv. Mater*, 2006, 18, 1109-1120.
[426] Hurst, SJ; Han, MS; Jean, AKRL; Mirkin, CA. *Anal. Chem.*, 2007, 79, 7201-7205.
[427] Han, MS; Jean, AKRL; Oh, BK; Heo, J; Mirkin, C. A. *Angew.Chem., Int. Ed*, 2006, 45, 1807-1810.
[428] Han, MS; Jean, AKRL. Mirkin, CA. *J. Am. Chem. Soc.*, 2006, 128, 4954-4955.
[429] Han, MS; Jean, AKRL; Mirkin, CA. *Anal. Chem.*, 2007, 79, 6037-6041.
[430] Lin, Z; Chen, X; Jia, T; Wang, X; Xie, Z; Oyama, M; Chen, X. *Anal.Chem.*, 2009, 81, 830-833.
[431] Lee, KY; Kim, DW; Heo, J; Kim, JS; Yang, JK; Cheong, GW; Han, SW. *Bull. Korean Chem. Soc.*, 2006, 27, 2081-2083.
[432] Wei, H; Li, B; Li, J; Wang, E; Dong, S.*Chem. Commun*, 2007, 3735-3737,
[433] Liu, J; Lu, Y. *Nature Protocols*, 2006, 1, 246-252.
[434] Liu, J; Lu, Y. *J. Am. Chem. Soc.*, 2003, 125, 6642-6643.
[435] Liu, J; Lu, Y. *J. Am. Chem. Soc.*, 2005, 127, 12677-12683.
[436] Gerard, L; Coté, Ryszard, M; Lec, Michael, V. Pishko, *IEEE Sensors J*, 2003, 3, 251-266.
[437] Liu, T; Tang, J; Jiang, L. *Biochem Biophys Res Commun*, 2004, 313, 3-7.
[438] Su, X; Chew, FT; Li, SFY. *Anal Sci.*, 2000, 16, 107-14.
[439] Jin, G; Li, W; Yu, S; Peng, Y; Kong, J. *Analyst*, 2008, 133, 1367-1372.
[440] Gamby, J; Lazerges, M; Pernelle, C; Perrot, H; Girault, HH; Tribollet, B. *Lab Chip*, 2007, 7, 1607-1609
[441] Lazerges, M; Perrot, H; Rabehagasoa, N; Antoineb, E; Compere, C. *Chem. Commun*, 2005, 6020-6022.
[442] Sergeyeva, T; Lavrik, N; Rachkov, A; Kazantseva, Z; Elskaya, A. *Biosens. Bioelectron*, 1998, 13, 159-369.

[443] Tahir, ZM; Alocilja, EC. *Biosens. Bioelectron*, 2003, 18, 813-819.

[444] Hnaiein, M; Hassen, WM; Abdelghani, A; Wirth, CF; Coste, J; Bessueille, F; Leonard, D; Renault, N. J. *Electrochem. Commun*, 2008, 10, 1152-1154.

[445] Liu, H; Yang, Y; Chen, P; Zhong, Z. *Biochem. Engin. J*, 2009, 45, 107-112.

[446] Zhang, Y; Tadigadapa, S. *Biosens. & Bioelectron*, 2004, 19, 1733-1743.

[447] Danielsson, B. *J. Biotechnol*, 1990, 15, 187-200.

[448] Bhambhani, A; Kumar, CV. *Microp. and Mesop. Mater*, 2008, 109, 223-232.

[449] Katz, E; Willner, I; Wang, J. *Electroanalysis*, 2004, 16, 19-44.

[450] Xiao, Y; Patolsky, F; Katz, E; Hainfeld, JF; Willner, I. *Science*, 2003, 299, 1877-1881.

[451] Wanekaya, AK; Chen, W; Myung, NV; Mulchandani, A. *Electroanalysis*, 2006, 18, 533-550.

[452] Wang, J. *Analyst*, 2005, 130, 421-426.

[453] Chopra, N; Gavalas, VG; Bachas, LG; Hinds, BJ; Bachas, LG. *Anal. Lett*, 2007, 40, 2067-2096.

[454] Vamvakaki, V; Fouskaki, M; Chaniotakis, N. *Anal. Lett*, 2007, 40, 2271-2287.

[455] He, P; Xu, Y; Fang, Y. *Microchim Acta*, 2006, 152, 175-186.

[456] Luo, X; Morrin, A; Killard, AJ; Smyth, MR. *Electroanalysis*, 2006, 18, 319-326.

[457] Wang, J. *Electroanalysis*, 2005, 17, 7-14.

[458] Pingarron, JM; Sedeno, PY; Cortes, AG. *Electrochim. Acta*, 2008, 53, 5848-5866.

[459] Malinauskas, A; Malinauskiene J; Ramanavicius, A. *Nanotechnology*, 2005, 16, R51-R62.

[460] Yun, YH; Dong, Z; Shanov, V; Heineman, WR; Halsall, HB; Bhattacharya, A; Conforti, L; Narayan, RK; Ball, WS; Schulz, M. J. *Nanotoday*, 2007, 2, 30-37.

[461] Yogeswaran, U; Chen, SM. *Sensors*, 2008, 8, 290-313.

[462] Rivas, GA; Rubianes, MD; Rodrıguez, MC; Ferreyra, NF; Luque, GL; Pedano, ML; Miscoria, SA; Parrado, C. *Talanta*, 2007, 74, 291-307.

[463] Guo, S; Dong, S. *Trends in Anal.Chem*., 2009, 28, 96-109.

[464] Astuti, Y; Palomares, E; Haque, SA; Durrant, JR. *J. Am. Chem. Soc*., 2005, 127, 15120-15126.

[465] Park, SJ; Taton, TA; Mirkin, CA. *Science*, 2002, 295, 1503-1506.

[466] Velev, OD; Kaler, EW. *Langmuir*, 1999, 15, 3693-3698.

[467] Zhaoa, ZW; Chenb, XJ; Tay, BK; Chenc, JS; Hana, ZJ; Khor, KA. *Biosens.& Bioelectron*, 2007, 23, 135-139.
[468] Choi, HN; Kim, MA; Lee, WY. *Anal. Chim. Acta*, 2005, 537, 179-187.
[469] Chu, X; Duan, D; Shen, G; Yu, R. *Talanta*, 2007, 71, 2040-2047.
[470] Yanga, Y; Yanga, H; Yanga, M; liua, Y; Shena, G; Yu, R. *Anal.Chim. Acta*, 2004, 525, 213-220.
[471] Li, Q; Luo, G; Feng, J; Zhou, Q; Zhang, L; Zhu, Y. *Electroanalysis*, 2001, 13, 413-416.
[472] Lin, J; Zhang, L; Zhang, S. *Anal. Biochem*, 2007, 370, 180-185.
[473] Xiao, Y; Ju, HX; Chen, HY. *Anal. Chim. Acta*, 1999, 391, 73.
[474] Musameh, M; Wang, J; Merkoci, A; Lin, YH. *Electrochem. Commun*, 2002, 4, 743-746.
[475] Xu, JJ; Zhao, W; Luo, XL; Chen, HY. *Chem. Commun*, 2005, 792-794.
[476] Wang, J; Xu, D; Kawde, AN; Polsky, R. *Anal. Chem*., 2001, 73, 5576-5581.
[477] Numnuam, A; Torres, KYC; Xiang, Y; Bash, R; Thavarungkul, P; Kanatharana, P; Pretsch, E; Wang, J; Bakker, E. *J. Am. Chem. Soc*., 2008, 130, 410-411.
[478] Numnuam, A; Torres, KYC; Xiang, Y; Bash, R; Thavarungkul, P; Kanatharana, P; Pretsch, E; Wang, J; Bakker, E. *Anal. Chem*., 2008, 80, 707-712.
[479] Torres, KYC; Dai, Z; Rubinova, N; Xiang, Y; Pretsch, E; Wang, J; Bakker, E. *J. Am. Chem. Soc*., 2006, 128, 13676-13677.
[480] Luo, XL; Xu, JJ; Zhao, W; Chen, HY. *Biosens. & Bioelectron*, 2004, 19, 1295-1300.
[481] Thurer, R; Vigassy, T; Hirayama, M; Wang, J; Bakker, E; Pretsch, E. *Anal. Chem*., 2007, 79, 5107-5110.
[482] Katz, E; Willner, I. *Electroanalysis*, 2003, 15, 913-947.
[483] KOwino, IO; Sadik, OA. *Electroanalysis*, 2005, 17, 2101-2113.
[484] Daniels, JS; Pourmand, N. *Electroanalysis*, 2007, 19, 1239-1257.
[485] Guan, JG; Miao, YQ; Zhang, QJ. *J. Biosc. Bioengin*, 2004, 97, 219-226.
[486] Miao, YQ; Guan, JG. *Anal. Lett*, 2004, 37, 1053-1062.
[487] Franks, W; Schenker, I; Schmutz, P; Hierlemann, A. *IEEE Transaction Biomed. Engineering*, 2005, 52, 1295-1302.
[488] Wilner, OI; Guidotti, C; Wieckowska, A; Gill, R; Willner, I. *Chem. Eur. J*, 2008, 14, 7774-7781.
[489] Yu, X; Lv, R; Ma, Z; Liu, Z; Hao, Y; Li, Q; Xu, D. *Analyst*, 2006, 131, 745-750.

[490] Alfonta, L; Zayats, IBM; Baraz, L; Kotler, M; Willner, I. *ChemBioChem*, 2004, 5, 949-957.

[491] Mantzila, AG; Prodromidis, MI. *Electroanalysis*, 2005, 17, 1878-1885.

[492] Kim, G; Mun, JH; Om, AS. *J. Phys., Conf. Series*, 2007, 61, 555-559.

[493] Rodriguez, MC; Kawde, AN; Wang, J. *Chem. Commun*, 2005, 4267-4269.

[494] Bardea, A; Patolsky, F; Dagan, A; Willner, I. *Chem. Commun.*, 1999, 21-22.

[495] Xu, Y; Ye, X; Yang, L; He, P; Fang,, Y. *Electroanalysis*, 2006, 18, 1471-1478.

[496] Khan, R; Dhayal, M. *Electrochem. Commun*, 2008, 10, 492-495.

[497] Chen, CP; Ganguly, A; Wang, CH; Hsu, CW; Chattopadhyay, S; Hsu, YK; Chang, YC; Chen, KH; Chen, LC. *Anal. Chem.*, 2009, 81, 36-42.

[498] Wang, TJ; Tu, CW; Liu, FK. *IEEE J. Selected Topics in Quantum Electron*, 2005, 11, 493-499.

[499] Yonzon, CR; Stuart, DA; Zhang, X; McFarland, AD; Haynes, CL; Duyne, RPV. *Talanta*, 2005, 67, 438-448.

[500] Hoa, XD; Kirk, AG; Tabrizian, M. *Biosens. & Bioelectron*, 2007, 23, 151-160.

[501] Hiep, HM; Endo, T; Kerman, K; Chikae, M; Kim, DK; Yamamura, S; Takamura, Y; Tamiya, E; *Sci.Technol. Adv. Mater*, 2007, 8, 331-338.

[502] Piliarik, M; Vaisocherov, H; Homola, J. *Biosens. & Bioelectron*, 2005, 20, 2104-2110.

[503] Lyon, LA; Musick, MD; Natan, MJ. *Anal. Chem.*, 1998, 70, 5177-5183.

[504] Lyon, LA; Musick, MD; Smith, PC; Reiss, BD; Pena, DJ; Natan, MJ. *Sens. Actuators B*, 1999, 54, 118-124.

[505] Ramanavièius, A; Herberg, FW; Hutschenreiter, S; Zimmermann, B; Lapenaite, I; Kausaite, A; Finkelsteinas, A; Ramanavieiene, A. *Acta Medica Lituanica*, 2005, 12, 1-9.

[506] Ladd, J; Taylor, AD; Piliarik, M; Homola, J; Jiang, S. *Anal. Chem.*, 2008, 80, 4231-4236.

[507] Kim, DK; Kerman, K; Saito, M; Sathuluri, RR; Endo, T; Yamamura, S; Kwon, YS; Tamiya, E. *Anal. Chem.*, 2007, 79, 1855-1864.

[508] Parramon, JS. *Nanotechnology*, 2009, 20, 235706.

[509] Liu, Y; Brandon, R; Cate, M; Peng, X; Stony R; Johnson, M. *Anal. Chem.*, 2007, 79, 8796-8802

[510] Sandros, MG; Shete V; Benson, DE. *Analyst*, 2006, 131, 229-235.

[511] Huang, CP; Li, YK; Chen, TM. *Biosens & Bioelectron*, 2007, 22, 1835-1838.

[512] Griffin, J; Singh, AK; Senapati, D; Rhodes, P; Mitchell, K; Robinson, B; Yu, E; Ray, PC. *Chem. Eur. J*, 2009, 15, 342-351.
[513] Shi, L; Rosenzweig, N; Rosenzweig, Z. *Anal. Chem*., 2007, 79, 208-214.
[514] Feng, CL; Zhong, X; Steinhart, M; Caminade, AM; Majoral, JP; Knoll, W. *Adv. Mater*, 2007, 19, 1933-1936.
[515] Tang, B; Cao, L; Xu, K; Zhuo, L; Ge, J; Li, Q; Yu, L. *Chem. Eur. J*, 2008, 14, 3637-3644.
[516] Gill, R; Bahshi, L; Freeman, R; Willner, I. *Angew. Chem. Int. Ed*, 2008, 47, 1676-1679.
[517] Dyadyusha, L; Yin, H; Jaiswal, S; Brown, T; Baumberg, JJ; Booy, FP; Melvin, T. *Chem. Commun*, 2005, 3201-3203.
[518] Duong, HD; Rhee, JI. *Talanta*, 2007, 73, 899-905.
[519] Yuan, J; Guo, W; Yin, J; Wang, E. *Talanta*, 2009, 77, 1858-1863.
[520] Shang, ZB; Wang, Y; Jin, WJ. *Talanta*, 2009, 78, 364-369.
[521] Bahshi, L; Freeman, R; Gill, R; Willner, I. *Small*, 2009, 5, 676-680.
[522] You, CC; Miranda, OR; Gider, B; Ghosh, PS; Kim, IB; Erdogan, B; Krovi, SA; Bunz, UHF; Rotello, VM. *Nat. Nanotechnol*, 2007, 2, 318-323.
[523] Miranda, OR; You, CC; Phillips, R; Kim, IB; Ghosh, PS; F. Bunz, UH; Rotello, VM. *J. Am. Chem. Soc*., 2007, 129, 9856-9857.
[524] Kneipp, J; Kneipp, H; McLaughlin, M; Brown, D; Kneipp, K. *Nano Lett*, 2006, 6, 2225-2231.
[525] Tripp, RA; Dluhy, RA; Zhao, Y. *Nanotoday*, 2008, 3, 31-37.
[526] Doering, WE; Piotti, ME; Natan, MJ; Freeman, RG. *Adv. Mater*, 2007, 19, 3100-3108.
[527] Kim, JH; Kang, T; Yoo, SM; Lee, SY; Kim, B; Choi, YK. *Nanotechnology*, 2009, 20, 235302.
[528] Merlen, A; Gadenne, V; Romann, J; Chevallier, V; Patrone, L; Valmalette, JC. *Nanotechnology*, 2009, 20, 215705.
[529] Deng, S; Fan, HM; Zhang, X; Loh, KP; Cheng, CL; Sow, CH; Foo, YL. *Nanotechnology*, 2009, 20, 175705.
[530] Zheng, X; Guo, D; Shao, Y; Jia, S; Xu, S; Zhao, B; Xu, W. *Langmuir*, 2008, 24, 4394-4398.
[531] Chen, J; Jiang, J; Gao, X; Liu, G; Shen, G; Yu, R. *Chem. Eur. J*, 2008, 14, 8374 - 8382.
[532] Hu, J; Zheng, PC; Jiang, JH; Shen, GL; Yu, RQ; Liu, GK. *Anal. Chem*., 2009, 81, 87-93.
[533] Kim, JH; Kim, JS; Choi, H; Lee, SM; Jun, BH; Yu, KN; Kuk, E; Kim, YK; Jeong, DH; Cho, MH; Lee, YS. *Anal. Chem*., 2006, 78, 6967-6973.

[534] Qian, X; Qian, X; Peng, XH; Ansari, D; Goen, QY; Chen, GZ; Shin, DM; Yang, L; Young, AN; Wang, MD; Nie, S. *Nature Biotechnol*, 2007, 26, 83-890.

[535] Kneipp, K; Haka, AS; Kneipp, H; Badizadegan, K; Yoshizawa, N; Boone, C; Peltier, KES; Motz, JT; Dasari, RR; Feld, MS. *Appl. Spectrosc*, 2002, 56, 150-154.

[536] Driskell, JD; Kwarta, KM; Lipert, RJ; Porter, MD; Neill, JD; Ridpath, JF. *Anal. Chem*., 2005, 77, 6147-6154.

[537] Aroca, RF; Puebla, RAA; Pieczonka, N; Cortez, SS; Ramos, JVG. *Adv. Colloid Interface Sci*., 2005, 116, 45-61.

[538] Cao, YC; Jin, R; Mirkin, CA. *Science*, 2002, 297, 1536-1540.

[539] Cao, YC; Jin, RC; Nam, JM; Thaxton, CS; Mirkin, CA. *J. Am. Chem. Soc*., 2003, 125, 14676-14677.

[540] Nam, JM; Stoeva, SI; Mirkin, CA. *J. Am. Chem. Soc*., 2004, 126, 5932-5933.

[541] Nam, JM; Thaxton, CS; Mirkin, CA. *Science*, 2003, 301, 1884-1886.

[542] Stoeva, SI; Lee, JS; Thaxton, CS; Mirkin, CA. *Angew. Chem. Int. Ed*, 2006, 45, 3303-3306.

[543] Liu, T; Liu, B; Zhang, H; Wang, Y. *J. Fluorescence*, 2005, 15, 729-733.

[544] Ye, Z; Tan, M; Wang, G; Yuan, J. *Anal. Chem*., 2004, 76, 513-518.

[545] Huhtinen, P; Kivela, M; Kuronen, O; Hagren, V; Takalo, H; Tenhu, H; Lovgren, T; Harma, H. *Anal. Chem*., 2005, 77, 2643-2648.

[546] Chen, Y; Chi, Y; Wen, H; Lu, Z. *Anal. Chem*., 2007, 79, 960-965.

[547] Ai, K; Zhang, B; Lu, L. *Angew. Chem. Int. Ed*, 2009, 48, 304-308.

[548] Sin, KK; Chan, CPY; Pang, TH; Seydack, M; Renneberg, R. *Anal. Bioanal. Chem*., 2006, 384, 638-644.

[549] Tan, M; Wang, G; Hai, X; Ye, Z; Yuan, J. *J. Mater. Chem*., 2004, 14 , 2896-2901.

[550] Yang, H; Santra, S; Walter, GA; Holloway, PH. *Adv. Mater*, 2006, 18, 2890-2894.

[551] Huang, CC; Su, CH; Li, WM; Liu, TY; Chen, JH; Yeh, CS. *Adv. Funct. Mater*, 2009, 19, 249-258.

[552] Louis, C; Bazzi, R; Marquette, CA; Bridot, JL; Roux, S; Ledoux, G; Mercier, B; Blum, L; Perriat, P; Tillement, O. *Chem. Mater*, 2005, 17, 1673-1682.

[553] Wang, C; Chen, J; Talavage, T; Irudayaraj, J. *Angew. Chem. Int. Ed*, 2009, 48, 2759-2763.

[554] Son, SJ; Bai, X; Lee, SB. *Drug Discovery Today*, 2007, 12, 657-663.

[555] Portney, NG; Ozkan, M. *Anal. Bioanal. Chem*., 2006, 384, 620-630.

[556] Margolis, DJA; Hoffman, JM; Herfkens, RJ; Jeffrey, RB; Quon, A; Gambhir, SS. *Radiology*, 2007, 245, 333-356.
[557] Weissleder, R. *Nat. Rev. Cancer*, 2001, 2, 11-18.
[558] Sharrna, P; Brown, S; Walter, G; Santra, S; Moudgil, B. *Adv. Colloid Interface Sci.*, 2006, 123, 471-485.
[559] Tan, WH; Wang, KM; He, XX; Zhao, XJ; Drake, T; Wang, L; Bagwe, RP. *Med. Res. Rev.*, 2004, 24, 621-638.
[560] Kircher, MF; Mahmood, U; King, RS; Weissleder, R; Josephson, L.*Cancer Res.*, 2003, 63, 8122-8125.
[561] Schellenberger, EA; Sosnovik, D; Weissleder, R; Josephson, L. *Bioconjug. Chem.*, 2004, 15, 1062.
[562] Jana, NR; Ying, JY. *Adv. Mater*, 2008, 20, 430-434.
[563] Thorek, DLJ; Chen, AK; Czpryna, J; Tsourkas, A. *Ann. Biomed. Engin.*, 2006, 34, 23-38.
[564] Rhyner, MN; Smith, AM; Cao, X; Mao, H; Yang, L; Nie, S. *Nanomedicine*, 2006, doi.10.2217/17435889.
[565] Goldman, ER; Medintz, IL; Mattoussi, H. *Anal. Bioanal. Chem.*, 2006, 384, 560-563.
[566] Shi, D; Cho, HS; Chen, Y; Xu, H; Gu, H; Lian, J; Wang, W; Liu, G; Huth, C; Wang, L; Ewing, RC; Budko, S; Pauletti, GM; Dong, Z. *Adv. Mater*, 2009, 21, 1-4.
[567] Barinov, A; Gregoratti, L; Dudin, P; Rosa, SL; Kiskinova, M. *Adv. Mater*, 2009, 21, 1-5.
[568] Levi, J; Cheng, Z; Gheysens, O; Patel, M; Chan, CT; Wang, Y; Namavari, M; Gambhir, SS. *Bioconjugat. Chem.*, 2007, 18, 628-634.
[569] Chao, JI; Perevedentseva, E; Chung, PH; Liu, KK; Cheng, CY; Chang, CC; Cheng, CL; *Biophys. J*, 2007, 93, 2199-2208.
[570] Zheng, Y; Gao, S; Ying, JY. *Adv. Mater*, 2007, 19, 376-380.
[571] Alivisatos, AP. *Science*, 1996, 271, 933-937.
[572] Smith, AM; Dave, S; Nie, S; True, L; Gao, X. *Expert Rev. Mol. Diag*, 2006, 6, 231-244.
[573] Bruchez, M; Moronne, M; Gin, P; Weiss, S; Alivisatos, AP. *Science*, 1998, 281, 2013-2016.
[574] Cao, YW; Banin, U. *J. Am. Chem. Soc.*, 2000, 122, 9692-9702.
[575] Mattoussi, H; Mauro, JM; Goldman, ER; Anderson, GP; Sundar, VC; Mikulec, FV; Bawendi, MG. *J. Am. Chem. Soc.*, 2000, 122, 12142-12150.
[576] Jaiswal, JK; Mattoussi, H; Mauro, JM; Simon, SM. *Nat. Biotechnol*, 2003, 21, 47-51.

[577] Dubertret, B; Skourides, P; Norris, DJ; Noireaux, V; Brivanlou, AH; Libchaber, A. *Science*, 2002, 298, 1759-1762.

[578] Derfus, AM; Chan, WCW; Bhatia, SN. *Nano Lett*, 2004, 4, 11-18.

[579] Hines, MA; Sionnest, PG. *J. Phys. Chem.*, 1996, 100, 468.

[580] Chan, WCW; Nie, SM. *Science*, 1998, 281, 2016-2018.

[581] Selvan, ST; Tan, TT; Ying, JY. *Adv. Mater*, 2005, 17, 1620-1625.

[582] Liu, HJ; Liu, J; Haley, KN; Treadway, JA; Larson, JP; Ge, NF; Peale, F; Bruchez, MP. Wu, XY. *Nat. Biotechnol*, 2003, 21, 41-46.

[583] Goldman, ER; Balighian, ED; Mattoussi, H; Kuno, MK; Mauro, JM; Tran, PT; Anderson, GP. *J. Am. Chem. Soc.*, 2002, 124, 6378-6382.

[584] Medintz, IL; Clapp, AR; Mattoussi, H; Goldman, ER; Fisher, B; Mauro, JM. *Nat. Mater*, 2003, 2, 630-.

[585] Choi, HS; Liu, W; Misra, P; Tanaka, E; Zimmer, JP; Ipe, BI; Bawendi, MG; Frangioni, JV. *Nat. Biotechnol*, 2007, 25, 1165-1170.

[586] So, MK; Xu, CJ; Loening, AM; Gambhir, SS; Rao, JH. *Nat. Biotechnol*, 2006, 24, 339-343.

[587] Zhou, M; Ghosh, I. *Biopolymers*, 2007, 88, 325-339.

[588] Chang, E; Miller, JS; Sun, JT; Yu, WW; Colvin, VL; Drezek, R; West, JL. *Biochem. Biophys. Res. Commun*, 2005, 334, 1317-1321.

[589] Jamieson, T; Bakhshi, R; Petrova, D; Pocock, R; Imani, M; Seifalian, AM. *Biomaterials*, 2007, 28, 4717-4732.

[590] Rao, JH; Andrasi, AD; Yao, HQ; Yao, HQ. *Curr. Opin. Biotechnol*, 2007, 18, 17-25.

[591] Velikov, KP; van Blaaderen, A. *Langmuir*, 2001, 17, 4779-4786.

[592] Qhobosheane, M; Zhang, P; Tan, WH. *J. Nanosci. Nanotechnol*, 2004, 4, 635-641.

[593] Yan, JL; Estevez, MC; Smith, JE; Wang, KM. He, XX; Wang, L; Tan, WH. *Nano Today*, 2007, 2, 44-50.

[594] Bagwe, RP; Yang, CY; Hilliard, LR; Tan, WH. *Langmuir*, 2004, 20, 8336-8342.

[595] Herr, JK; Smith, JE; Medley, CD; Shangguan, DH; Tan, WH. *Anal. Chem.*, 2006, 78, 2918-2924.

[596] Zheng, XL; Liu, Y; Pan, M; L, XQ; Zhang, JY; Zhao, CY; Tong, YX; Su, CY. *Angew. Chem. Int. Ed*, 2007, 46, 7399-7403.

[597] Shan, J; Chen, J; Meng, J; Collins, J; Soboyejo, W; Friedberg, JS; Ju, Y. *J. Appl. Phys.*, 2008, 104, 094308-6.

[598] Harma, H; Keranen, AM; Lovgren, T. *Nanotechnology*, 2007, 18, 075604.

[599] Schellenberger, E; Schnorr, J; Reutelingsperger, C; Ungethm, L; Meyer, W; Taupitz, M; Hamm, B. *Small*, 2008, 4, 225-230.
[600] Zhao, M; Beauregard, DA; Loizou, L; Davletov, B; Brindle, KM. *Nat. Med*, 2001, 7, 1241-1244.
[601] Lee, J; Yang, J; Ko, H; Oh, SJ; Kang, J; Son, JH; Lee, K; Lee, SW; Yoon, HG; Suh, JS; Huh, YM; Haam, S. *Adv. Funct. Mater*, 2008, 18, 258-264.
[602] Shin, J; Anisur, RM; Ko, MK; Im, GH; Lee, JH; Lee, IS. *Angew. Chem. Int. Ed*, 2009, 48, 321 -324.
[603] Hu, F; Wei, L; Zhou, Z; Ran, Y; Li, Z; Gao, M. *Adv. Mater*, 2006, 18, 2553-2556.
[604] Sun, C; Lee, JSH; Zhang, M. *Adv. Drug Delivery Rev.*, 2008, 60, 1252-1265.
[605] Chertok, B; Moffat, BA; David, AE; Yu, F; Bergemann, C; Ross, BD; Yang, VC. *Biomaterials*, 2008, 29, 487-496.
[606] Na, HB; Song, IC; Hyeon, T. *Adv. Mater*, 2009, 21, 1-16.
[607] Xu, ZP; Kurniawan, ND; Bartlett, PF; Lu, GQ. *Chem. Eur. J*, 2007, 13, 2824-2830.
[608] Insin, N; Tracy, JB; Lee, H; Zimmer, JP; Westervelt, RM; Bawendi, MG. *ACS Nano*, 2008, 2, 197-202.
[609] Wunderbaldinger, P; Josephson, L; Weissleder, R. *Acad. Radiology*, 2002, 9, S304-S306.
[610] Wunderbaldinger, P; Josephson, L; Weissleder, R. *Bioconjugate Chem.*, 2002, 13, 264-268.
[611] Weissleder, R; Kelly, K; Sun, EY; Shtatland, T; Josephson, L. *Nat. Biotechnol*, 2005, 23, 1418-1423.
[612] Sun, EY; Josephson, L; Weissleder, R. *Mol. Imaging*, 2006, 5, 122-128.
[613] Sosnovik, DE; Nahrendorf, M; Weissleder, R. *Circulation*, 2007, 115, 2076-2086.
[614] Fawell, S; Seery, J; Daikh, Y; Moore, C; Chen, LL; Pepinsky, B; Barsoum, J. *Proc. Natl. Acad. Sci.* USA, 1994, 91, 664-668.
[615] Lewin, M; Carlesso, N; Tung, CH; Tang, XW; Cory, D; Scadden, T; Weissleder, R. *Nat. Biotechnol*, 2000, 18, 410-414.
[616] Weissleder, R; Moore, A; Mahmood, U; Bhorade, R; Benveniste, H; Chiocca, E; Basilion, JB. *Nat. Med*, 2000, 6, 351-355.
[617] Enochs, WS; Harsh, G; Hochberg, F; Weissleder, R. *J. Magn. Reson. Imag.*, 1999, 9, 228-232.
[618] Contag, PR; Olomu, IN; Stevenson, DK; Contag, CH. *Nat. Med*, 1998, 4, 245-247.

[619] Zhao, M; Beauregard, DA; Loizou, L; Davletov, B; Brindle, KM. *Nat. Med*, 2001, 7, 1241-1244.

[620] Thompson, CB. *Science*, 1995, 267, 1456-1462.

[621] Poptani, H; Puumalainen, A; Grohn, OHJ; Loimas, S; Kainulainen, R; Yla-Herttuala, S; Kauppinen, RA. *Cancer Gene Ther.*, 1998, 5, 101-109.

[622] Francis, G; Blankenberg, P; Katsikis, D; Storrs, RW; Beaulieu, C; Spielman, D; Chen, JY; Naumovski, L; Tait, JF. *Blood*, 1997, 89, 3778-86.

[623] Nunn, AVW; Barnard, ML; Bhakoo, K; Murray, J; Chilvers, EJ; Bell, JD. *FEBS Lett*, 1996, 392, 295-298.

[624] Mornet, S; Vasseur, S; Grasset, F; Duguet, E. *J. Mater. Chem.*, 2004, 14, 2161-2175.

[625] Hauck, TS; Ghazani, AA; Chan, WCW. *Small*, 2008, 4, 153-159.

[626] Ghosh, P; Han, G; De, M; Kim, CK; Rotello, VM. *Adv. Drug Delivery Rev.*, 2008, 60, 1307-1315.

[627] Soppimath, KS; Liu, LH; Seow, WY; Liu, SQ; Powell, R; Chan, P; Yang, YY. *Adv. Funct. Mater*, 2007, 17, 355-362.

[628] Brigger, I; Dubernet, C; Couvreur, P. *Adv. Drug Delivery Rev.*, 2002, 54, 631-651.

[629] Bhattacharya, R; Mukherjee, P. *Adv. Drug Delivery Rev.*, 2008, 60, 1289-1306.

[630] Gu, FX; Karnik, R; Wang, AZ; Alexis, F; Nissenbaum, EL; Hong, S; Langer, RS; Farokhzad, OC. *Nanotoday*, 2007, 2, 14-21.

[631] Kumar, MN., Mohapatra, SS; Kong, X; Jena, PK; Bakowsky, U; Lehr, CM. *J. Nanosci. Nanotechnol*, 2004, 4, 990-994.

[632] Gelperina, S; Kisich, K; Iseman, MD; Heifets, L. *Am. J. Respir. Crit. Care Med*, 2005, 172, 1487-1490.

[633] Derfus, AM; Maltzahn, G; Harris, TJ; Duza, T; Vecchio, KS; Ruoslahti, E; Bhatia, SN. *Adv. Mater*, 2007, 19, 3932-3936.

[634] Zhang, J; Misra, RDK. *Acta Biomat*, 2007, 3, 838-850.

[635] Polizzi, MA; Stasko, NA; Schoenfisch, MH. *Langmuir*, 2007, 23, 4938-4943.

[636] Neuman, D; Ostrowski, AD; Absalonson, RO; Strouse, GF; Ford, PC. *J. Am. Chem. Soc.*, 2007, 129, 4146-4147.

[637] O'Neal, DP; Hirsch, LR; Halas, NJ; Payne, JD; West, JL.*Cancer Lett*, 2004, 209, 171-176.

[638] Perkel, JM. *TheScientist*, 2004, 18, 14-18.

[639] DeNardo, SJ; DeNardo, GL; Miers, LA; Natarajan, A; Foreman, AR; Gruettner, C. *Clin. Cancer Res.*, 2005, 11, 7087s-7092s.

[640] El-Sayed, IH; Huang, X; El-Sayed, MA. *Cancer Lett*, 2006, 239, 129-135.

[641] Kaul, G; Amiji, M. *Pharm. Res.*, 2005, 22, 951-961.

INDEX

A

absorption spectroscopy, 45, 49
accommodation, 13
acid, 21, 64, 82
acidity, 21, 22
activation energy, 20, 60
active site, 69
adenine, 67
adhesion, 27, 64
adhesion properties, 27
adsorption, 20, 21, 64, 72, 82, 85
aerogels, 31
AFM, 1, 45, 55
aggregates, 65
aggregation, 64
aggregation process, 64
alcohol, 31, 69
alloys, 11, 32, 36, 43
aluminum, 22, 30
aluminum oxide, 22
ammonium, 33, 35, 37
amplitude, 59, 70
anisotropy, 18
annealing, 11, 12
antibody, 64, 66, 72, 75, 83
anticancer drug, 82
antigen, 62, 66, 67, 68, 72, 75
anti-inflammatory drugs, 82
antimony, 68
apoptosis, 81
applications, 1, 2, 3, 6, 7, 9, 14, 18, 27, 30, 45, 61, 64, 65, 66, 67, 72, 77, 79, 85
argon, 27
arthritis, 82
assumptions, 22
atomic distances, 48
atomic force, 55
atomic nucleus, 54
atoms, 3, 15, 19, 20, 27, 31, 41, 46, 47, 48, 50, 51, 52
attachment, 67, 73
attacks, 82
attractiveness, 36
Au nanoparticles, 82
authors, 80, 81
availability, 80

B

background, 78
backscattering, 54
bacteria, 72, 74
band gap, 9, 10, 16
barriers, 10, 15
basicity, 22
baths, 32
beams, 29
behavior, 3, 13, 18, 22, 56, 62, 63, 70, 80
benign, 34

benzene, 33
bias, 60
binding, 21, 49, 64, 65, 73, 74, 81, 82, 83
binding energies, 21
binding energy, 49
bioassay, 77
bioavailability, 81
biocompatibility, 63, 64, 69, 78, 85
biological systems, 55
biomarkers, 63, 78
biomaterials, 61, 72, 78
biosensors, 2, 61, 62, 63, 64, 65, 66, 67, 68, 69, 70, 72, 73, 85
biotechnology, 62, 76
bleaching, 77, 78
blood, 78, 81
blood vessels, 78
bonding, 21, 49, 64
bonds, 22
bone, 80
bone marrow, 80
boundary surface, 60
breast cancer, 62
buffer, 22, 82
bulk materials, 1, 9, 45
burn, 83
by-products, 41

C

cancer, 62, 75, 78, 83
cancer cells, 62, 75, 78
candidates, 61
carbon, 29, 34, 37, 38, 39, 68
carbon nanotubes, 29, 34, 37, 39, 68
carrier, 82
catalysis, 2, 19, 25, 63, 64, 78
catalyst, 20, 22, 23, 28, 32, 34, 35, 62
catalytic activity, 19, 22
catalytic properties, 19, 23, 49, 67, 72
catalytic system, 67
category a, 35
cation, 64
cell, 3, 14, 48, 70, 72, 76, 77, 80, 81, 82
cell membranes, 81
cell surface, 80
ceramic, 14, 31, 82
cerium, 15, 22, 68
channels, 3, 37, 39
chemical bonds, 21, 56
chemical properties, 7, 76
chemical reactivity, 63
chemical stability, 64
chemical vapor deposition, 25, 28, 29
chemisorption, 23
chemotherapy, 82
cholesterol, 68
circulation, 82
clarity, 53
classes, 5
classification, 3, 85
clusters, 20
CMC, 38
CO_2, 56
coatings, 31
Collage, 100, 101
color, 18, 64, 65
communication, 68, 69
community, 34
competition, 74
complementary DNA, 75
complexity, 72
compliance, 81
components, 36, 50, 55, 60
composites, 35, 42
composition, 1, 3, 28, 30, 35, 36, 45, 49, 51, 57, 61
compounds, 15, 51, 59, 73
computed tomography, 75
concentration, 15, 22, 32, 36, 38, 40, 57, 59, 68, 69, 70, 82
condensation, 29, 31
conductance, 3, 66
conduction, 8, 16
conductivity, 14, 15, 59, 66, 67
conductor, 15, 60
configuration, 66
confinement, 8, 15, 16, 63, 68
Congress, iv
conjugation, 78, 81, 82

conservation, 16
consolidation, 25
construction, 69, 70, 72, 74, 80
contamination, 11, 22, 41, 56
control, 1, 25, 28, 29, 30, 33, 35, 36, 61, 80, 81
conversion, 2, 78
coordination, 22, 49
Copyright, iv
coupling, 58, 78
creep, 10
critical value, 17
crystal growth, 29, 33
crystal structure, 3, 43, 45, 46, 47
crystalline, 3, 28, 29, 35, 46, 47, 49, 55
crystallinity, 38, 45
crystallites, 48, 55
crystallization, 37, 40
crystals, 7, 28, 34
curing, vii, 1
CVD, 25, 26, 28, 29, 30
cyst, 69
cytometry, 81
cytoplasm, 62

D

damages, iv
data processing, 61
database, 47
decay, 79
defects, 14, 30, 45, 50, 54, 57
definition, 48
deformation, 13, 42
degradation, 80
delivery, 63, 81, 83, 85
density, 13, 21, 34, 41, 66, 79
deposition, 25, 26, 27, 28, 29, 30, 37, 41, 68
deposition rate, 27, 29
desorption, 20
detection, 57, 62, 63, 64, 65, 66, 67, 68, 70, 72, 74, 75, 76, 78, 81
deviation, 47, 48
dielectric constant, 9
dielectrics, 75
differential scanning, 67
differentiation, 80
diffraction, 46, 47, 48, 53, 54, 58
diffusion, 13, 20, 41, 60
diffusion process, 13
diffusion rates, 20
diffusivity, 22, 34
dimensionality, 5
direct observation, 52
discipline, vii, 1
discrimination, 74
diseases, 65
dislocation, 10, 11, 13
dispersion, 31
distribution, 7, 35, 36, 48, 51, 54, 60, 81
diversity, 27, 30, 34, 61
DNA, 2, 37, 61, 62, 65, 68, 70, 74, 75, 82, 85
dopants, 14
doping, 14, 15, 82
drug delivery, 2, 61, 62, 81, 85
drug release, 81
drugs, 81, 82
drying, 32
ductility, 10, 13
dyes, 58, 65, 75, 76, 78
dynamics, vii, 1, 104

E

electric charge, 14
electric field, 42
electrical conductivity, 15, 85
electrical properties, 68, 72
electrochemical deposition, 37
electrochemistry, 25, 63, 65, 68
electrodes, 2, 41, 59, 67, 69, 70, 72
electrolyte, 41, 59
electromagnetic, 7, 46, 54, 82
electron, 2, 7, 9, 15, 16, 21, 26, 29, 45, 49, 50, 51, 52, 53, 54, 57, 63, 64, 67, 68, 69, 73
electron beam lithography, 2
electron diffraction, 29, 45, 54
electron microscopy, 52, 54

electron paramagnetic resonance, 45
electronic structure, 8
electrons, 14, 15, 21, 29, 48, 49, 50, 51, 52, 53, 54
electrophoresis, 42
emission, 2, 57, 58, 62, 63, 73, 75, 77
emulsions, 36
encapsulation, 78, 82
energy, 2, 7, 9, 10, 11, 15, 16, 18, 20, 22, 26, 27, 29, 36, 42, 43, 45, 49, 50, 51, 54, 56, 57, 65, 73, 78, 82
energy transfer, 65, 73, 78, 82
engineering, 2
enlargement, 70
environment, 36, 69, 82
environmental control, 68
enzyme immobilization, 68, 69
enzymes, 57, 64, 67, 69, 72, 76, 78
epitaxial films, 28
equilibrium, 11, 73, 104
equipment, 68, 72
ESR, 50
ethanol, 69
evacuation, 27
evaporation, 26, 28, 29
evolution, 31, 68
EXAFS, 49
excitation, 16, 50, 58, 73, 77, 78
exciton, 7, 9
extraction, 37, 77

F

fabrication, 3, 25, 28, 31, 61, 64, 65, 85
FAD, 67
family, 48, 49, 55
FCC, 43
Fermi level, 21
fibers, 31, 62
films, 27, 28, 29, 30, 38
filtration, 35
fluctuations, 48
fluorescence, 58, 63, 65, 73, 74, 75, 76, 78
fluorophores, 58, 77, 78
focusing, 27
Ford, 115
free energy, 36
free radicals, 50
FTIR, 56
fuel, 3, 14, 22, 68
functionalization, 61, 65, 77

G

gallium, 27
gases, 27, 28, 49
gel, 26, 31, 41, 78
gelation, 31
gene, 61, 80, 83
gene expression, 80
gene therapy, 80
generation, 14, 66
genes, 80
genetic defect, 80
glasses, 31
glucose, 66, 67, 68, 69
glucose oxidase, 67, 69
glutamate, 78
glutathione, 37, 82
glycol, 2, 33, 83
glycoproteins, 74
gold, 33, 56, 62, 65, 67, 68, 70, 72, 73, 74, 75, 77, 82, 83
gold nanoparticles, 33, 65, 67, 68, 74, 75, 78, 82
grain boundaries, 10, 11, 15, 42, 59
grains, 36, 42
graphite, 70
gravity, 31
groups, 22, 38, 56, 74, 77
growth, 11, 13, 28, 32, 33, 34, 35, 37, 38, 40, 41, 83
growth factor, 83

H

halos, 55
hardness, 10, 11, 13
health, 83, 85

health effects, 85
heat, 15, 20, 21, 26, 27, 83
heating, 26, 82
hemoglobin, 66
hepatitis, 67
HRTEM, 1, 45
hybrid, 2, 67, 69, 82
hybridization, 62, 70, 75
hydrazine, 36
hydrides, 41
hydrocarbons, 56
hydrogen, 22, 43, 64, 68, 69, 79, 82
hydrogen bonds, 82
hydrogen peroxide, 69
hydrolysis, 31, 33, 37
hydrophobicity, 63
hydroquinone, 68
hydrothermal synthesis, 34
hydroxide, 35, 41
hydroxyl, 30, 31
hydroxyl groups, 30

I

ideal, 66
identification, 45, 46, 56, 74
illumination, 55
image, 54, 58, 61, 79, 80
images, 45, 51, 53, 54, 55
imaging modalities, 79
immobilization, 63, 64, 73
impurities, 43, 57
in vitro, 76, 77, 81, 82, 85
in vivo, 75, 76, 77, 80, 81, 82, 85
incidence, 47, 72
indentation, 56
induction, 26
industry, 1, 29, 34, 68, 83
inelastic, 54
infinite, 18
instruments, 53, 65, 72
insulators, 17
interaction, 29, 50, 64, 72, 73, 75, 78
interactions, 16, 17, 19, 64, 68, 74, 75
interface, 18, 23, 38, 68, 72
interference, 50, 69
intermetallic compounds, 10, 42
intermetallics, 43
internalizing, 80
intravenously, 80, 82
ionization, 20, 21, 50
ions, 14, 15, 21, 27, 31, 33, 41, 65
iron, 11, 18, 20, 62, 68, 79, 82, 83
irradiation, 33
isomerization, 23

K

kinetics, 72

L

labeling, 2, 62, 76, 79
lactase, 68
lactate dehydrogenase, 78
lanthanide, 77, 79
laser ablation, 28
lasers, 7
lattice parameters, 55
lattices, 48
lens, 54
Lewis acids, 31
life sciences, 76
lifetime, 49, 57
ligand, 31, 77
light conditions, 78
light scattering, 79
light transmission, 52
limitation, 27, 60
line, 30, 57, 80
liposomes, 79, 82
liquid phase, 31, 36, 37
low temperatures, 27, 29
luciferase, 78
luminescence, 15, 16, 77, 78
Luo, 88, 94, 97, 107, 108
lymph, 78
lymph node, 78
lysozyme, 37

M

macromolecules, 57
magnetic field, 27, 50, 79
magnetic moment, 17, 50
magnetic particles, 18, 80
magnetic properties, 9, 17, 18
magnetic resonance, 2, 75, 81
magnetic resonance imaging, 2, 75
magnetic resonance spectroscopy, 81
magnetization, 17, 18
majority, 59, 66
manganese, 68
manipulation, 3
manufacturing, 86
marrow, 80
material surface, 56
materials science, 46
matrix, 68, 69, 73
measurement, 50, 56, 59, 70
mechanical properties, 10, 56, 61
media, 36
melting, 7, 11, 20
melting temperature, 11
membranes, 31, 64
mercury, 58
mesoporous materials, 52
metal hydroxides, 35
metal nanoparticles, 2, 32, 36, 37, 61, 69, 72
metal oxides, 2, 9, 10, 42, 56, 64
metal salts, 31, 36
metals, 3, 11, 13, 14, 21, 36, 42, 43, 49, 79
metastatic cancer, 82
methodology, 28, 65, 71
mice, 80, 83
microemulsion, 39, 78
micrometer, 5, 7, 55
microscope, 52, 53, 55, 58
microscopy, 55, 81
microspheres, 33, 79
microstructure, 42, 54
microwave radiation, 50
microwaves, 50
migration, 22
miniaturization, 1
mixing, 30, 35, 57
mobility, 22
MOCVD, 26, 28, 29
model, 48
modelling, 86
models, 80
modulus, 56
molecular beam, 28, 29
molecular beam epitaxy, 28, 29
molecules, 19, 21, 32, 36, 37, 38, 40, 56, 57, 58, 64, 65, 72, 74, 76
monolayer, 69
monomers, 42
Moon, 91, 103
morphology, 22, 23, 25, 85
motion, 10, 14
motivation, 1
movement, 8, 15
MRI, 2, 75, 79, 80

N

NaCl, 35
NAD, 69
NADH, 69
nanobelts, 5, 85
nanocomposites, 42, 63
nanocrystalline metals, 11
nanocrystals, 3, 7, 9, 15, 16, 25, 32, 78
nanodevices, vii, 1
nanoelectronics, 2
nanofibers, 2
nanomaterials, 2, 3, 5, 7, 8, 32, 33, 37, 38, 39, 45, 56, 61, 62, 64, 68, 85
nanomedicine, vii, 1
nanometer, 1, 3, 7, 15, 37, 39, 42, 54
nanometer scale, 1, 7, 37, 39
nanometers, 36, 42, 54
nanoparticles, 2, 5, 16, 25, 32, 33, 34, 35, 36, 37, 39, 41, 45, 49, 56, 62, 63, 64, 65, 66, 67, 68, 74, 75, 76, 77, 79, 81, 82, 83, 85
nanophases, 35
nanophotonics, 2
nanorods, 32, 33, 34, 38, 40, 45, 85

nanoscale structures, 61
nanostructured materials, vii, 1, 2, 3, 25, 26, 28, 31, 37, 45, 53, 56, 61, 64, 69, 70, 72, 74, 77, 78, 85
nanostructures, 1, 5, 6, 8, 25, 34, 39, 41, 45, 56, 62, 85
nanotechnology, vii, 1, 45, 83, 85, 88
nanotube, 38, 39, 85
nanowires, 2, 3, 5, 29, 33, 34, 37, 38, 40, 45, 85
necrosis, 83
Netherlands, 100, 101, 104
network, 31
neurodegenerative diseases, 81
niobium, 68
NIR, 78, 83
nitric oxide, 82
NMR, 79
noble metals, 33
nuclear magnetic resonance, 79
nucleation, 28, 34, 35, 41, 42
nuclei, 35
nucleic acid, 64, 72, 75
nucleus, 62

O

observations, 51
oil, 36, 39
one dimension, 2, 5, 8, 25
opacity, 77
opportunities, vii, 1, 86
optical properties, 7, 45, 85
optoelectronic properties, 25
order, 14, 16, 22, 27, 36, 47, 54, 70, 71, 80
organ, 81
orientation, 3, 48
oxalate, 34
oxidation, 18, 49, 59, 69, 77
oxide nanoparticles, 33, 36, 62, 63, 68, 79
oxides, 9, 14, 15, 18, 19, 21, 22, 29, 31, 35, 38
oxygen, 14, 15, 22, 30, 60, 64, 85
oxygen sensors, 14

P

PAA, 38
palladium, 37
paradigm, 74
parallel, 20, 36, 50, 55, 59
parameter, 5, 48
parameters, 3, 52, 60
partial differential equations, 71
particle morphology, 19, 23
particles, 11, 14, 15, 16, 17, 18, 19, 22, 31, 34, 35, 36, 38, 42, 43, 51, 54, 61, 65, 66, 67, 73, 78, 79, 80, 82
pathogenesis, 80
pathogens, 74
PCR, 65, 75
pepsin, 66
peptides, 61, 78
performance, 72
permission, iv
PET, 75
phospholipids, 81
photoabsorption, 49
photobleaching, 79
photocatalysts, 33
photoluminescence, 9, 57
photons, 49, 53
physical properties, 1, 7, 82
piezoelectricity, 64
PL spectrum, 57
Planck constant, 50
plasma, 27, 28, 29, 81
plasma membrane, 81
plastic deformation, 13
platinum, 37, 68
point mutation, 68
polarity, 41, 59
polarization, 65
polycarbonate, 37
polycondensation, 37
polymer, 7, 16, 32, 37, 38, 66, 73, 74, 75, 77, 82
polymer matrix, 16, 66
polymerase, 65
polymerase chain reaction, 65

polymeric materials, 2, 33
polymerization, 31
polymers, 64, 66
porosity, 11, 13
porous materials, 37, 39
positron, 75
positron emission tomography, 75
potassium, 56
power, 14, 27, 34
precipitation, 26, 31, 34, 35
pressure, 26, 28, 33, 34, 47, 49, 60
probability, 20
probe, 46, 55, 56, 65, 78
production, 30, 37, 49
project, 58
proliferation, 80
promoter, 80
properties, 1, 2, 3, 5, 6, 7, 8, 9, 13, 15, 17, 19, 23, 31, 45, 46, 49, 50, 55, 56, 57, 59, 61, 64, 66, 69, 72, 85, 104
proteins, 57, 61, 62, 63, 64, 65, 74, 75, 76, 78, 79, 80, 82
proteolysis, 78
proteolytic enzyme, 78
protocol, 65
protons, 21, 79
prototype, 27
pulse, 28, 79
pumps, 49
pure water, 33

Q

quantum confinement, 7, 8
quantum dot, 2, 3, 7, 25, 29, 56, 61, 62, 63, 77, 78, 82
quantum dots, 2, 3, 7, 25, 29, 56, 61, 62, 63, 77, 78, 82
quantum well, 29
quartz, 66

R

radiation, 32, 46, 49, 56, 57, 79
radio, 26, 27, 76
radius, 7, 9, 21, 35
Ramadan, 93
Raman spectra, 74
Raman spectroscopy, 74
range, 1, 14, 16, 20, 28, 36, 37, 42, 43, 49, 56, 57, 59, 65, 66, 67, 71, 73, 77, 78, 85
reactant, 19, 35, 64, 69
reaction rate, 59
reaction time, 41
reactions, 19, 20, 21, 22, 23, 28, 30, 31, 32, 33, 41, 47, 57, 64, 67, 72
reactivity, 21
reagents, 76
real time, 47
reception, 66
receptors, 64, 72, 73
recognition, 62, 64, 68, 74, 85
recommendations, iv
recovery, 79
red shift, 16
reflectivity, 72
refractive index, 72, 73
region, 56, 83
rejection, 81
relationship, 55, 69
relaxation, 11, 57, 60, 79
relaxation process, 60
relaxation processes, 60
resistance, 60, 72, 79
resolution, 45, 47, 53, 54, 55, 56, 80, 81
respect, 16, 35, 86
restructuring, 21
returns, 73
reusability, 63
rights, iv
RNA, 65
rods, 25, 61
Romania, 32
room temperature, 26
root-mean-square, 48
roughness, 21

S

safety, 78, 81
salt, 35, 68
salts, 31, 35, 36
sampling, 51
saturation, 18
scanning electron microscopy, 45
scattering, 47, 48, 54, 63, 74
segregation, 15
selected area electron diffraction, 45
selectivity, 19, 23, 63, 64, 67, 68
self-assembly, 37, 40
self-destruction, 81
semiconductor, 2, 7, 9, 10, 29, 56, 61, 67, 77
semiconductors, 3, 16, 62, 64, 69, 74
sensing, 61, 62, 63, 65, 67, 73, 74
sensitivity, 21, 57, 63, 64, 66, 67, 68, 70, 72, 74
sensors, 25, 65, 66, 67, 70, 74
separation, 37, 55, 63, 72, 75, 78
shape, 13, 17, 25, 33, 37, 51, 59, 60, 61, 72
shear, 42
side effects, 81
signal transduction, 62
signals, 53, 59, 62, 71, 73, 80
signal-to-noise ratio, 79
silica, 37, 63, 68, 73, 77, 79
silicon, 16, 22, 29, 55
silver, 56, 62, 68, 70, 77
single crystals, 34, 36
single walled carbon nanotubes, 2
sintering, 35
SiO_2, 2, 19, 38, 63, 68, 75
SiO_2 surface, 68
sodium, 33, 35
sodium hydroxide, 35
solar cells, 7
sol-gel, 31, 32
solid oxide fuel cells, 14
solid state, 46
solid tumors, 83
solubility, 33, 77
solvents, 37, 78
solvothermal synthesis, 34
space, 16, 54, 55, 79
species, 22, 30, 31, 35, 50, 56, 57, 59, 62, 69, 75
specific surface, 2
spectroscopy, 45, 49, 50, 56, 70, 74
spectrum, 7, 14, 16, 46, 50, 60, 73, 74, 77, 79
speed, 33
spin, 17, 50
stability, 11, 57, 63, 68, 77, 78, 85
standards, 51
STM, 1, 55
stoichiometry, 15, 28, 47
storage, 43, 61, 64, 82, 85
strain, 13, 16, 54
strategies, 61, 77, 80
strategy, 80
strength, 10, 13, 16, 21, 50
stress, 13
stretching, 56
structural changes, 57
substrates, 28, 30, 37, 38, 39, 67
Sun, 89, 90, 91, 92, 93, 94, 95, 96, 98, 99, 102, 104, 113, 114
superplasticity, 13
supply, 27
surface area, 2, 7, 22, 61, 64, 68, 73, 85
surface chemistry, 30, 82
surface modification, 78
surface properties, 3
surface region, 49
surface structure, 19
surfactant, 33, 36, 37, 38, 40
survival, 83
SWNTs, 39
symmetry, 5
synthesis, 2, 3, 6, 11, 25, 26, 29, 30, 32, 33, 34, 37, 38, 39, 42, 43, 65, 78

T

targets, 75
TEM, 45, 52, 53, 54

temperature, 11, 13, 14, 18, 23, 28, 31, 33, 35, 41, 47, 60, 67, 82
tension, 13
TEOS, 37
testing, 80
texture, 51
therapeutics, 61
therapy, 83
thermal energy, 18
thermal evaporation, 26, 28
thermal stability, 35
thermodynamics, 15
thin films, 2, 5, 29, 30, 32, 45, 46
tin, 68
tin oxide, 68
tissue, 2, 76, 77, 79
titanium, 33, 41, 68
titanium isopropoxide, 33
toxicity, 64, 77, 81, 85
transducer, 62, 64, 70
transduction, 64, 66
transferrin, 80, 83
transgene, 80
transition, 9, 16, 21, 22, 31, 50, 57
transition metal, 21, 22
transitions, 8, 47, 58
transmission, 45, 52, 53, 66
Transmission Electron Microscopy (TEM), 45, 54
transparency, 16
transport, 3, 7, 9, 15, 34, 63
tuberculosis, 82
tumor, 83
tumor growth, 83
tumors, 83
tungsten, 27
turnover, 68

U

ultrasound, 32, 75
uniform, 38, 55
urea, 66, 68

V

vacancies, 14, 15, 47
vacuum, 19, 21, 26, 27, 28, 29, 49
valence, 8, 16, 73
vapor, 25, 26, 28, 37, 38
vector, 16, 48
viruses, 74
viscosity, 34

W

wave vector, 16
wavelengths, 15
weight ratio, 22
wires, 25, 61
workers, 33, 41, 74

X

XPS, 45, 49
X-ray diffraction (XRD), 45, 46, 54
X-ray photoelectron spectroscopy (XPS), 49
XRD, 1

Y

yttrium, 35

Z

zinc, 38, 68
zinc oxide, 68
zirconium, 68
ZnO, 2, 7, 10, 14, 27, 28, 29, 30, 32, 33, 34, 38, 39, 42, 59, 63, 66, 68
ZnO nanorods, 29, 34, 39
ZnO nanostructures, 27